M000288429

Successful Wine Marketing

Edited by

Kirby Moulton, PhD
Economist, Emeritus
Department of Agricultural and Resource Economics
University of California, Berkeley
Berkeley, California

James Lapsley, PhD
Chair, Department of Science,
Agriculture and Natural Resources
University Extension
University of California, Davis
Davis, California

 Springer

Library of Congress Cataloging-in-Publication Data

Successful wine marketing / edited by Kirby Moulton, James Lapsley.
p. cm.
Includes index.
ISBN 0-8342-1962-X
1. Wine industry.   2. Wine—Marketing.   I. Moulton, Kirby S.   II. Lapsley, James T.
HD9370.5.S83 2001
663'.2'00688—dc21
00-068241

© 2001 Springer Science+Business Media, Inc., originally published by Aspen Publishers, Inc.

Printed in the United States of America.

springeronline.com

Editorial Services: Erin McKindley
Library of Congress Catalog Number: 00-068241
ISBN: 0-8342-1962-X

Printed in the United States of America

2 3 4 5

Table of Contents

Contributors

Steve Boone
Chairman
Boone Turrini Inc.
Orinda, California

Michel Bourqui, MBA, DEA
Administrator and Marketing
  Coordinator
Office International de la Vigne
  et du Vin
Paris, France

James Cahill, MS
Vice President of Sales, North
  America
Supreme Corq Inc.
Kent, Washington

Ralph Colonna
President
CF/NAPA
Napa, California

J. Daniel Davis
Senior Attorney
Wine, Beer & Spirits Law Group
Pillsbury Winthrop LLP
San Francisco, California

J. Patrick Dore
Vice President of Marketing and
  Sales
ASV Wines
San Francisco, California

Tom Eddy
President
Thomas G. Eddy and Associates
Wine Maker
Tom Eddy Winery
Calistoga, California

Ed Everett
Past President (Deceased)
New World Wines
San Francisco, California

Alex G. Franco
Sales Operations Manager
The Clorox Sales Co.
Oakland, California

Steve Fredricks
Vice President and Partner
Turrentine Wine Brokerage
San Anselmo, California

Jon A. Fredrikson, MBA
President
Gomberg, Fredrikson & Associates
San Francisco, California

Richard A. Gooner, MBA, PhD
Assistant Professor of Marketing
University of Alabama
Tuscaloosa, Alabama

Peter Granoff
Founder and Master Sommelier
wine.com
Napa, California

Stephanie Grubbs
Proprietor
Moss Lane Wine Marketing
Napa, California

Michael R. Hagerty, PhD
Professor
Graduate School of Management
University of California
Davis, California

Michael C. Houlihan
President and CEO
Grape Links Inc. and Barefoot
  Cellars
Santa Rosa, California

**Agustin Francisco Huneeus,
  MBA**
President
Franciscan Estates
St. Helena, California

Leslie Litwak, MBA
Vice President, Marketing
Seagram Chateau & Estates Wine
  Company
Rutherford, California

Christian M. Miller, MBA
Director
MKF Research
Motto, Kryla & Fisher LLP
St. Helena, California

Fred Myers, MA
Vice President and Eastern
  Division Manager
Frederick Wildman & Sons Ltd.
New York, New York

Michael B. Newman, JD
Resident Partner
Buchman & O'Brien
San Francisco, California

Bruce H. Rector
Wine Maker
Abundance Vineyards
Glen Ellen, California

Robert D. Reynolds, MS
President
Reynolds Economics
Moraga, California

James M. Seff, JD
Partner in Charge
Wine, Beer & Spirits Law Group
Pillsbury Winthrop LLP
San Francisco, California

Thomas H. Shelton
President and CEO
Joseph Phelps Vineyards
St. Helena, California

John Skupny
Proprietor
Lang & Reed Wine Company and
    Vineyard Avenue Marketing
Vice President of Business
    Development, uvine.com
St. Helena, California

Anthony L. Spawton
International Director
The Wine Marketing Research
    Group
University of South Australia
South Australia, Australia

Preface

This book reflects the work of wine marketing experts as expressed in their presentations to the annual three-week Wine Marketing Short Course at the University of California, Davis. The course was initially organized in collaboration with the international wine management curriculum sponsored by the International Organization for Vines and Wines (OIV). We have been involved in this course since its inception a decade ago.

This book is intended for students in wine marketing and management, enology, and viticulture who seek to broaden their understanding of the wine sector. It is also intended for those already working in wine marketing and management who seek new ideas and insights. Finally, this book should be of general interest to others involved directly or indirectly in the grape and wine sector.

Each chapter was written from the oral presentations of the authors and reflects the spontaneity and informality of the classroom environment. The writing may lack the "gravitas" of academic material, but it accurately presents the thinking and conclusions of those who make a living by marketing wine. There is some duplication that serves to emphasize important points, and there are several case studies explaining real-life experiences in the industry. Legal requirements and commercial practices cited by authors may differ between regions and among institutions familiar to readers. However, the underlying principles guiding marketing strategies can be applied in different situations, for example, where supermarket wine sales may be restricted or direct sales prohibited. The book is not intended as a data source. Each author presents data, in many cases without attribution, which are relevant to the chapter topic and to the time the initial class presentation was made. These data are generally consistent between chapters but may not be precisely so.

The message that comes through is that there is no "cookbook" approach to wine marketing. Successful wine marketing rather depends on the creative application of well-established principles to individual marketing opportunities. These principles can be applied regardless of political boundaries, consumer behavior or market size; they result in strategies that are appropriate for any situation.

Acknowledgments

This book would not have been possible without the continuing support that the authors have given to the annual Wine Marketing Course at the University of California, Davis. Their efforts have provided new insights to wine marketing for a wide variety of industry members, domestic and international students, and other motivated persons. We particularly appreciate their patience and cooperation during the period over which their individual chapters have been edited.

We would also like to specifically acknowledge the early leadership efforts provided by Professor Carole Meredith, Department of Viticulture and Enology at the University of California, Davis. She was instrumental in organizing and contributing to the class. Her early efforts and the continued support of the Department provided the professional environment needed to attract the authors represented in this book. We are pleased to acknowledge the initiatives of Michel Bourqui, Administrator in the International Office of Vines and Wines (OIV). As administrator of the international masters level course in wine marketing and management, under the auspices of OIV, he provided the financial core of international students that allowed the program to grow at Davis and led to the publication of this book.

We are especially grateful for the contributions of Geralyn Unterberg, Editor in the Department of Agricultural and Resource Economics, University of California, Berkeley, who provided her professional skills in transcribing the initial classroom presentations by the authors, reworking the edited copies, and maintaining the liaisons and files needed for a project of this scope.

CHAPTER 1

Introduction: Consumer Behavior and Marketing Strategies

Kirby Moulton, Anthony L. Spawton, and Michel Bourqui

INTRODUCTION

This chapter links universal marketing principles such as market segmentation, product differentiation, and distribution management to consumer behavior and regional marketing targets. The same principles may suggest that a given small winery should concentrate on local restaurants serving Italian cuisine and that a large producer should adapt a multiple-brand strategy aimed at international markets. Different locations may have consumers with the same characteristics and require similar marketing strategies. Where marketing needs are different, the marketing mix must be altered to meet those needs irrespective of geographic boundaries. Thus, market segments may be regional or transnational in scope. Marketing strategies are dictated primarily by the nature of targeted market segments. This suggests that *disaggregation*, not *globalization*, is the appropriate term to describe what is happening in wine marketing today.

The wine market is beset by change. Trade barriers have diminished, investment capital flows freely around the world, communication is almost instantaneous, and consumer behavior has evolved from traditional to

---

Anthony L. Spawton is international director and lecturer at The Wine Marketing Research Group, University of South Australia, South Australia. He helped develop the first Web-based, University-accredited wine marketing course and established the curriculum for the OIV program in Australia.

Michel Bourqui is administrator and marketing coordinator within the Office International de la Vigne et du Vin (OIV). He is the former director, and one of the founders, of the OIV graduate course in wine management and marketing.

1

experimental. Producers and governments concerned with marketing have had to adjust production, distribution, and political strategies to cope with these changes. This chapter summarizes some of the factors affecting consumer behavior and their implications for marketing strategies by industry or government.

INFLUENCING CONSUMER BEHAVIOR

Complex factors influence consumer behavior toward wine. Together, these factors shape geographic, demographic, and use-defined markets. Most of the factors are demographic, concerning consumers' location, income, culture, education, and age. Producers have little influence over these external factors. But producers can influence "internal" factors, such as information about wine and its consumption. Advertising, promotion (including packaging, place of sale, and display), and media are used most often to influence these internal factors. Such marketing strategies aim to trigger consumer purchases by convincing consumers that a particular product satisfies their needs or by encouraging consumers to change one or more of those needs. Needs include the consumer's status, risk acceptance, convenience, and taste—desires to be exclusive or in the lead, desires for lower risk in purchase and consumption, or desires for "good" taste. Marketing programs seek to build consumer awareness of the product, loyalty to the product, perceptions of the product's quality, and favorable images of the product.

Perhaps the greatest shift in consumer purchase behavior over the past generation has been from table wines to quality wines, particularly in traditional wine-drinking countries. Preferences among quality wines have changed as consumers in some areas have shifted toward wines with varietal designations, while those in other areas have shifted toward wines with regional or vineyard designations. The increased consumption of higher-quality wine was prompted, in part, by increased knowledge about the health effects of wine. And the rapid spread of international cuisines has brought with it an interest in different wines to accompany these "new" foods. As a result, consumer taste expectations have increased. New consumers, replacing the "baby boomers" of the past, appear to be less influenced by tradition, reputation, and wine experts. They are hedonistic rather than traditional.

CONSUMER CLASSIFICATION

Relatively few people consume wine in many countries. In the United States, for example, perhaps 40 or 45 percent of the adult population does not drink any alcoholic beverages. Of the remainder, less than 20 percent are "regular" wine drinkers, and they account for most of the wine consumed. The most interesting category is that of occasional wine drinkers; it has double the population of the "regular" wine drinker category. The occasional wine drinkers tend to like wine and might buy more of it if persuaded to do so.

The following estimate of consumer classes is based on a review of studies in numerous markets. The figures represent our judgment of what seems reasonable for understanding how markets are structured.

1. *Connoisseurs* account for perhaps 5 percent of wine consumers. They are knowledgeable about wine and demanding in their requirements. They are quality conscious, but some are egocentric, are prejudiced, and "look down at the others."
2. *Aspirants* account for perhaps 45 percent of wine consumers. They definitely want to know more about wine. They tend to be curious, open-minded, intellectually honest, and ready to experiment. They may suffer from an inferiority complex relative to those more familiar with wine.
3. *Newcomers* are not very interested in wine and do not drink much of it. They tend to take whatever advice or product is given to them. They represent probably 35 percent of wine consumers.
4. *Simple drinkers* consume wine by habit or custom but have no particular interest in it. They may represent 15 percent of wine consumers. If the first three consumer classes are more or less evenly spread among world consumers, this class is mainly composed of older people in traditional wine-producing countries.

On first analysis this appears to be a sad situation. Only 5 percent of consumers are really knowledgeable about the product, and the remainder may be nice but need considerable education. The average consumer in most other nonwine markets faces far fewer product choices and, consequently, tends to be better informed about them and more "rational" in their selection. The uneasy, confused feeling of many wine consumers could become tiring, if not dissuasive, to wine sellers and governments with marketing programs.

On second analysis, however, this classification scheme indicates that there is opportunity for growth. Producers are facing a market where 80 percent of the wine consumers are waiting for an efficient and relevant message. How many other markets enjoy such potential? Producers need to develop a sincere, intelligent, and qualitative relationship between the product and the consumer. There is no point in launching a costly, innovative product-development scheme until the consumer knows what the product attributes are and the producer knows what the consumer preferences are.

MARKETING STRATEGIES

The needs of the market have a clear impact on the design of the marketing mix. The product has to have a set of attributes that are meaningful to the consumer and can be related to consumer needs. Such attributes must "personally" answer the following questions: What is the wine made of? Where is it made? By whom is it made? The attributes should communicate "personally" because they feed a personal relationship between the producer and the consumer. The place of encounter between the consumer and the product has to be adequately arranged to help the "relationship" develop solidly. The price must be relevant to the product's value, as perceived by a given consumer. A thinking, honest consumer will not pay more than what he or she considers the product to be worth.

The attributes of the product should convey some sense of the level of risk undertaken by the purchaser. New wines produced by longtime producers may convey less risk. New producers producing in recognized appellations may also convey less risk. (Appellations are districts of grape production that generally, but not always, include certain rules concerning varieties, production practices, and yield, which produce wines with identifiable characteristics.) Proper communication will document the range of risk from the "no-risk-taken" weak relationship (standard brand for standard product) to the greater "discovery" risk, with its highly rewarding relationship (new brand for new product).

Communication must be relevant to the place of sale as well as to product identity. Trade marketing is fundamental to all wine marketing strategies because the producer is dealing first with the "trade"—that is, distributors, retailers, and restaurants—and then with consumers. A "home-made" marketing strategy has little effect if it ignores the marketing concerns of the trade. Strategies that are effective in some markets incorporate jointly developed distributor agreements and category management for retailers.

Needs and Attributes

The first step in developing a marketing strategy is to find out what the consumer wants from a wine and how the seller can satisfy those desires. This may be done through elaborate market research projects using surveys, focus groups, and available market data. It may also be undertaken through trade interviews and personal observations. The results will help in calculating the degree to which appellation, variety, and brand influence purchase decisions for particular products in specific markets. The results will assist in positioning the product in the market and in determining how to support the product through pricing, advertising, and promotion. This process of market evaluation is termed a "market audit" and is the forerunner of a marketing plan.

A related step is to determine the relevant product attributes—that is, product characteristics that are meaningful to consumers. Chapter 3 identifies four major classes of attributes: quality, price, convenience, and signaling (elements that suggest other product characteristics). These attributes help direct consumer choices among competing products. For some products, one mix of attributes will be important, while for other products, a different mix will apply. The producer needs to determine which attributes are important in motivating customers to choose its products.

Product Differentiation

Product differentiation includes all the strategic decisions that seek to distinguish the product from competing products, to enhance its value and the revenue it generates. The goal is to make the wine a "product" and not a "commodity." Only adding value in the consumer's eyes can do this. Many strategies relating to product style, packaging, and distribution lose their impact over time because they are copied by competitors. Hence, firms need to change their product differentiation strategies continually. In this respect, differences based on terroir, appellation, and, to a lesser extent, brand are easier to sustain.

Terroir, for example, provides a solid basis for differentiation. However, there are other factors that affect how one wine is distinguished from another. These include the price bracket, brand name, value for cost, promotion activities, producer reputation and personality, awards, endorsement by publicists, wine education, outlets where sold, and quality of sales representation. Because of the complex interaction of factors influencing consumer perceptions of quality, any single term on a wine label may or may not be effective, depending on the market and the product.

Market Segmentation

Marketing strategies must be based on a clear understanding of the markets to be served. Wine marketers recognize three basic types of markets: the mass market, the product-differentiated market, and the niche (or target) market. The mass-marketing strategy provides one offering for all customers. This is the strategy of Coca-Cola. For appellation wines, an example of mass marketing would be a basic set of wines with an individual appellation sold in all markets. The advantage of this strategy is in the economies of scale gained in production and promotion. The disadvantage is the inability to meet the needs of important submarkets.

A product-differentiated marketing strategy offers two or more products with different features to the same target market. Thus, appellation wine makers might offer different wine qualities or different packaging to the same market. The principal advantage is the building up of brand loyalty in the market. The main disadvantage is that the region (or wine maker) is dependent entirely on a single market.

The target-market strategy is based on the ability to identify and evaluate product attributes and consumer perceptions that define specific market segments. The objective is to find clusters of consumers that have more similarities among themselves than they do with consumers outside the cluster. To be attractive, a market segment needs to be responsive to a common marketing strategy. Each segment has to be large enough to be financially attractive, it has to be accessible, and the strategy needs to be operational. By positioning wines in such niche markets, brand managers should be building the market value of their brand. The concepts of positioning and repositioning wine in various markets are critically important to wine marketers. They provide the incentive to investigate markets thoroughly and evaluate alternative positions. They also provide the key to understanding the principles of segmentation (see Chapters 19–22).

New Product Development

Beverage consumers have become accustomed to change as new products appear and older ones disappear with regularity. The development of flavored wines is an example of this. New product development among quality wines is more subtle. It usually involves a change in blends or fermentation and aging techniques that alter flavor. It may involve new labels or packages. But rarely, if ever, are quality wines changed to produce a truly new product. The strategy with quality wines is really one of product differentiation among competing wines. Differentiation based on terroir is one such strategy, although it does not produce a "new" product but rather seeks a better position for existing products. Appellation and terroir wine makers cannot rely on designing the wine to meet consumer needs but instead need to focus on how to combine the other market-mix factors in a way that stimulates consumer purchases.

Sales and Distribution

Making distribution work effectively is difficult. The wine market is structured in such a way that the distribution networks have enormous power relative to the usually smaller-scale quality-wine producers. Consequently, it helps to have a distributor policy that articulates the wine maker's philosophy about the wine and expectations for it. Collaboration with distributors is critical because the channel in which wine is sold affects its price, its presentation, perceptions of it, and its relationship to competing products.

The place of sale will influence consumer perceptions of the wine. The decision as to outlet must be made primarily on the basis of where the targeted consumers want to buy the wine. Appellation and terroir wines might be better positioned in specialty wine stores, where salespersons can explain the wines' good qualities, rather than in supermarkets, where such sales assistance is not provided. Sellers can influence the location of wines in retail stores but must make a special effort to do so.

Wine enterprises are beginning to use the Internet to sell product or advise others of the availability of wines or related materials. The Internet has become an important promotional tool, one that wine producers can no longer ignore. A Web site can help resolve such problems as the inequity in size between producers and distributors, the loss of wine expertise at the retail level, and the geographic exclusivity of many distribution systems (see Chapter 5).

CONCLUSIONS

The marketing strategies reviewed here have relevance throughout the world. Their implementation depends on the characteristics of target markets and the creativity of marketing organizations. Whether a seller is engaged in "guerilla marketing" to appeal to defined small-market segments or is active in large-scale global marketing, the same principles apply. This is aptly demonstrated in other chapters in this book. It is important in this regard to keep a clear distinction between marketing, which is the development and implementation of principles, and selling, which is the logistics of marketing and will differ according to local conditions.

Long-term efforts to expand the U.S. wine market are likely to be most successful if focused on existing consumers who like wine but do not drink much of it. Surveys and focus groups provide the data that will help identify the characteristics of the target audience and the type of messages to which they are likely to respond. Based on survey results in the United States, the message should be one that demystifies wine and broadens consumers' sense of the occasions when wine should be consumed.

Changes in wine marketing strategies have immediate and important impacts on the regulations and market policies of regional organizations, national governments, and international agreements. The increased emphasis on export markets raises issues of trade barriers, property rights, and mutual recognition that were quietly accepted before. Product differentiation creates pressure on regulations concerning permissible grape varieties and blending practices. Wine sellers observe that consumer behavior patterns spill across geographic regulatory borders and strongly press for similar regulations in different jurisdictions. This is particularly true for marketing within the United States, where a great variety of regulations exists. The tremendous capacity of the Internet to facilitate direct wine marketing is putting pressure on regulations that tend to protect older distribution systems.

Regional associations, trade groups, and governments involved in wine marketing programs can improve program effectiveness if they are aware of changed consumer behavior and new marketing strategies. This will require research to identify changes and select the best program strategies. Marketing programs should be based on strong plans that include production, financing, and organizational considerations. The plans should articulate appropriate terroir, brand, and variety strategies for each level in the wine industry. The plans also need macro- and microeconomic components, including concepts for joining with other producers or associations to expand the demand for target wines in domestic and export markets. The objective is to make a profit by convincing consumers that they will benefit more, even if they pay higher prices for specified wines than for competing wines. The appeal may be through a geographic image, such as that of terroir; through an appeal to a particular market niche using a brand; or through a modification in varieties or blending that produces a new and desired product. When wine demand increases, all producers have the opportunity to benefit.

PART I

Setting Sights on the Market

CHAPTER 2

Researching
the Wine Consumer

Christian M. Miller

INTRODUCTION

This chapter is about determining the characteristics of wine consumers. The purpose is to show how these determinations are made and how they can be used in wine marketing. The empirical focus is on the American consumer, but the methodology and implications are transferable to other markets. The first principle guiding consumer research and marketing strategy is that there is neither just one type of consumer nor an average consumer. If someone refers to *the* wine consumer as having this or that characteristic, the information is relatively useless. This chapter is less a comprehensive survey of the different types of wine consumers than an approach and an attitude toward market research and data interpretation. Using this approach, a winery should be able to define the characteristics of potential buyers for its products and what marketing strategies may trigger consumer purchases. The data quoted in this chapter without attribution have been summarized from company research results as well as from various surveys and reports available to the company in 1997. These include IRI and Nielsen household panel data, the Gomberg–Fredrikson Report, and Merrill Research–Wine Marketing Council Report

Christian M. Miller received his MBA from Cornell University in 1985 and has worked as a negotiant and in marketing for Kendall-Jackson and Sebastiani Vineyards. He works for Motto, Kryla & Fisher LLP, one of California's leading wine consulting firms, where he heads their syndicated research database projects.

(1997). The data indicate past market conditions, but are useful for illustrating the types and approximate sizes of various market segments that are the target of market action.

CLASSIFYING WINE CONSUMERS

The traditional view of wine consumers is a mixture of myth and fact. It postulates that a consumer "progresses" from table wines (according to European definitions), or jug and generic wines (according to American definitions), to "easy-to-drink" blends such as white zinfandel, to less expensive varietal and DOC (designation of controlled origin—used in Italy) wines, to superpremium chardonnay and AOC (appellation of controlled origin—used in France) wines; finally, the consumer evolves into a homo-vinifera, drinking nothing but luxury wines. Some people follow this consumption pattern. But most American wine drinkers do not.

A different view of wine consumers examines consumption of various wine types in the context of income and the volume of wine consumed. Figure 2–1 illustrates this idea.

The figure is based on U.S. experience but is hypothetical and does not present actual data. It shows that there is overlap among many buyers. It illustrates that infrequent wine consumers seem to prefer white zinfandel and low-priced varietals. Consumers in the same income bracket that tend to drink larger quantities of wine seem to prefer varietals in 1.5-liter sizes. Superpremium wines, not surprisingly, are the choice of high-income consumers with above-average consumption rates. The figure also shows that there is a group called the "bag-in-the-box consumer" that is mid to low income. Members of this group drink more wine than average and pay less for it. Their purchase criterion is the cost per glass, bottle, or box for a wine of standard quality. The bag in the box enables them to drink more at a reasonable cost.

The purchasers of superpremium wines may be wine connoisseurs. It is often assumed that someone buying a Kendall-Jackson Reserve or similar wine is by definition a wine aficionado. However, the person may only have a high income. The person may have heard that Kendall-Jackson is a fancier, higher-quality wine and buy it the few times that he or she buys any wine at all. It fits in with the person's lifestyle of luxury cars, fancy restaurants, and large homes.

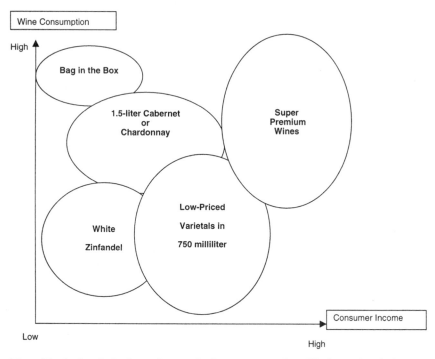

Note: Vertical axis is the volume of wine consumption. Horizontal axis is consumer income.

Figure 2–1 The Income-Consumption Matrix.

Figure 2–1 shows that consumers at various income levels may buy wines from several different price segments depending on the occasion and other personal preferences. Comparing price per bottle with market size also shows this. A surprising market has evolved in the United States, with the low-cost "jug" wines no longer outselling higher-cost "premium" wines. Jug wines are no longer the base of a wine consumption pyramid. Low-cost bag-in-the-box and generic wines, mostly in bottles of 1.5 liters or more, account for less than one-half of California table wine shipments. Some buyers of these wines also purchase wines in higher price classes, such as popularly priced varietal wines. At the other end of the scale are the buyers of ultrapremium wines; they are classified as either traditional or experimental buyers. This market segment also contains a few buyers of the somewhat less expensive superpremium wines. A similar phenomenon is observed in all price segments: there is a primary core of buyers that is augmented by a smaller number of people who consume wine infrequently but purchase primarily in either lower- or higher-priced segments.

Most wine consumers buy an assortment of different wines, brands, and sizes. Typically, they have one wine that they return to and consider to be their "house wine." However, for other purchases they do not have much brand loyalty. Even for their house wine they are willing to switch among several acceptable producers. Over time, these preferences can change. Research has shown that many consumers have moved from one category to another, both up and down the price scale. Significant movement has occurred into the standard premium and superpremium categories, and this movement has fueled sharp increases in wine revenues without a major change in volume. Consumers are drinking better wines and paying more for them. There is no doubt that most people buying more expensive wines start out with lower-priced wines and then move up to the superpremium wines, whether they drink a lot of wine or relatively little. This does not mean, however, that everyone started out at the lowest price and quality level.

WINE AND CONSUMER DEMOGRAPHICS

Demographically driven marketing is very popular now. However, many of the wines people buy are selected according to taste or image differences. Demographic data on income, age, education, profession, and gender may explain how a person can afford certain types of wines or how he or she may react to some promotional messages. But these data cannot predict taste and image preferences. This is difficult for a traditional marketer to understand. For example, marginal, core, and supercore wine buyers are classified by their wine consumption. The differences in their income, education, and ethnicity are not large. Demographic differences are not sufficient to explain differences in purchase behavior between the core and marginal groups.

Over the years, our company has used survey and focus group results, and has found that they show significant differences in each group's knowledge and approach to buying and enjoying wine. These sources show, for example, that 47 percent of marginal buyers drink wine only on special occasions, while just 9 percent of the supercore buyers are so restricted;

that wine would be the first beverage choice for informal occasions such as picnics by 20 percent of the core buyers but only 8 percent of the marginal buyers; and that wine is used for dinners at home by 73 percent of super-core buyers but only 44 percent of marginal buyers. Recognition of these sorts of behavioral differences will lead to marketing strategies more effective than those based only on traditional demographic relationships.

The good news for wine marketers is that it is possible to satisfy sales objectives with a relatively small number of consumers. The bad news is that a relatively large number of sellers are searching for similar market segments, and conditions are very competitive. The obvious challenge is to find that group of consumers that escapes the eye of most marketers. That is why direct marketing, the Internet, wine clubs, and mailing lists are of great interest to wineries, especially small ones. Such resources tap into a private, small slice of an already very small segment.

Taste often drives the market as much as income. A common myth is that chardonnay drinkers have a rich lifestyle and that white zinfandel is brought in a lunch bucket to a construction site by blue-collar workers. Marketers that accept this perception and base sales strategies on it are foolish. It turns out that there is considerable overlap at the same income levels for buyers of white zinfandel, average cabernet and chardonnay, and low-priced varietals. There may be 5 percent less white zinfandel sold at the higher end of the relevant income range and 5 percent more sold in the mid-range. One would expect similar results for drinkers of fruit-flavored wines. Higher income is likely to motivate consumers to buy better wines but not necessarily a greater volume. However, for those persons who have a taste for wine and are involved in a wine lifestyle (wine country tours, tastings, wine clubs, and the like), a higher income may well result in more purchases.

Stage of life often affects wine consumption. For example, people earning between $15,000 to $25,000 a year tend to consume more wine than those in the next income bracket. These drinkers tend to be young adults. As their income and age increase, they tend to take on family responsibilities that limit the number of special occasions during which wine is drunk. There is a negative correlation between wine consumption and the presence of several young children in the household. As this group gets older and their income continues to increase, their consumption starts to increase again.

Compared to the overall adult population, wine buyers tend to have higher incomes, to be older heads of household, and to work at white-collar jobs.

Relative to the average wine buyer, buyers of generic wines tend to be of middle income, to work in blue-collar occupations, and to have less education. Varietal wine buyers, typically chardonnay or cabernet sauvignon consumers, have higher incomes than the average wine buyer, with more education and younger heads of households. Cabernet buyers are more likely to be single and white. Pinot noir buyers show unusual loyalty to this wine. It is the only high-priced red varietal to show constant repurchase rates above 40 percent. The white zinfandel buyer tends to be of middle income, to have a small household, and to be a young or single head of household. This group of buyers tends to include more black and Hispanic consumers.

THE WHITE ZINFANDEL EXAMPLE

White zinfandel attracts consumers that prefer the taste of a wine that is somewhat sweet and slightly tart. It is a taste preference that seems to endure among many of these consumers. Results from surveys and household panels suggest that the white zinfandel market consists of first-time wine buyers, marginal consumers, and core consumers. White zinfandel in 750-milliliter bottles attracts the highest rate of first-time wine buyers among current wines. Consequently, white zinfandel producers are the most dependent on new wine buyers for growth in the business. White zinfandel is also more dependent than other varieties on the marginal wine buyer; 60 percent of its volume is derived from consumers who buy less frequently than once a month. Such buyers like the wine but are not involved in wine culture. They will pick up the wine for a birthday at home or to have as an alternative at a cocktail party. They are more likely to buy white zinfandel than many other types of wine.

The core buyers are responsible for a lot of the 1.5-liter volume. They are more likely to have traded up from generics and lower-priced white zinfandel. The core buyers consume more than average consumers and prefer the bargain of the larger package, which allows them to consume more without spending much more money. These consumers tend to stick with white zinfandel over time and are unlikely to trade up toward drier wines.

THE BOX WINE MARKET AND PACKAGE LOYALTY

The box wine market is an interesting niche market. It consists primarily of consumers with above-average consumption rates and strong preferences for value and convenience. Box wines sold in some supermarkets for $8.59 per five-liter box in 1998, far below the average price per liter for all generic wines. The market grew rapidly, at the expense of other sizes, but gained a very low household penetration rate. Only 3.9 percent of households bought these wines, according to a 1997 survey, one of the lowest penetration rates of any of the type-size combinations. The volume purchased per buyer is greater than for any other wine size combination, but the unit value is lower. Buyers have a high loyalty to that particular size and type, and most of the new business is derived from switching by generic buyers and increased consumption by current box buyers.

Loyalty to package size also occurs with wine in bottles. Household panel data show that only 17 percent of all buyers of 750-milliter and 1.5-liter bottles buy both sizes. The remaining 83 percent stick to one size or the other. Generally, regular wine drinkers are a higher proportion of 1.5-liter buyers than they are of 750-milliliter buyers.

The box format is unlikely to draw in many new buyers to the wine business despite the practicalities of the format. However, it might steal market share from traditional suppliers in the premium wine market. If an established quality brand were to come out packaged in more expensive boxes, that could change the dynamics of the market because it is a practical format for the regular wine drinker. The box is 2.5 inches wide and slides into a corner of the refrigerator. The box keeps wine fresh and available for the regular drinker at a comparatively low price.

TYPES OF BUYING DECISIONS AND FACTORS INFLUENCING THEM

Consumer purchase decisions may be planned down to the exact item to be purchased or to a particular variety, may be motivated by a particular reminder in a store, or may be strictly impulse. In each case, there are variable sets of factors that influence each of the decisions. The issue for the marketer is to determine at what point to approach this purchase process. The following list illustrates the relationship between decision types and influencing factors:

- planned purchase

  1. example: customer intention to buy brand X chardonnay for a party that night

    2. influences: brand ads, retailer ads, brand loyalty, varietal prefer-
ence

- substitute purchase

    1. example: customer realization that brand B chardonnay is on dis-
play, is cheaper than brand X, and looks good
    2. influences: package and price, in-store promotion, out-of-stocks,
coupons

- planned category

    1. example: customer plan to purchase white zinfandel but uncer-
tainty about which one to pick
    2. influences: brand image/awareness, price, shelf position, brand ads

- reminder

    1. example: wine display at meat counter that reminds customer to
buy red wine
    2. influences: in-store promotion, point of purchase (POP) material,
brand ads

- add-on

    1. example: shopper who needs wine for recipe and to serve with the
meal
    2. influence: occasion-driven usage, cross-merchandising

- impulse

    1. example: customer that tastes a pinot noir while shopping and
likes the wine
    2. influence: in-store promotions, POP materials

For the wine business as a whole, in-store decision making is probably more common than planned purchasing. This benefits wines that do not or cannot advertise and build brand awareness. It emphasizes the need for "buy me" labeling and packaging, good shelf and store location, and creative POP materials. Of course, publicity that increases brand awareness, and pricing that is consistent with the category, facilitate the in-store purchase decision. In-store merchandising is less important for planned purchases because the buyer already has been influenced through experience, advertising, or recommendations to seek out specific wines. For example, a winery should not expect the package to affect the planned purchase unless it is featured in visually driven advertising.

THE SOURCES OF SALES DATA

Market strategies are based on market intelligence. The more credible the intelligence, the better the strategy. Intelligence on market movement comes from three primary data sources: winery shipments, wholesaler depletions, and retail scanning results. Shipment data cover what wineries sell and ship. They are derived from tax reports because wineries must report and pay excise taxes on wines that are shipped from their bonded warehouses. These data are relatively easy and inexpensive to collect, particularly for one's own winery. The disadvantage is that shipments are not the same as sales to consumers and therefore will lag in defining trends. Typically, there is a one- to four-month lag between trends at retail and changes in winery shipments, and a two-month delay in reporting those shipments. Thus it is possible that a marketing manager might be six months behind in perceiving a retail trend. The published data are aggregated by company and do not include information on brands, varieties, or prices. The California State Board of Equalization is the primary source of shipment data for California, but usually the data are collected, collated, and published by private enterprises such as Gomberg-Fredrikson or Wine-Stats. The U.S. Department of Commerce reports export shipments. Trade associations and other groups publish quarterly and annual summaries.

Depletions are shipments from wholesalers to retailers, restaurants, and other customers. Most wholesalers will report depletions to suppliers. The advantage of depletions data over shipment data is that depletions are one step closer to ultimate buyers and therefore have less time lag. The disadvantage is that the data are usually available only for the supplying winery, without information about competitors and wine categories. Often, there is no breakdown by account or price.

Grocery stores and many other retailers now scan Universal Product Codes to record sales and price. Companies such as Nielsen and IRI purchase and organize these data and resell them to suppliers and wholesalers. This is an excellent source of information because it measures actual retail sales and gives prices and quantities for periods as short as one week. When combined with audits of advertising and promotion, retail scanning information is a powerful tool for analyzing sales (see Chapter 4). It provides a big advantage through its ability to organize data in large segments (such as all varietal wines) and small segments (such as brand A chardonnay in 750-milliliter bottles). It also can sum up the data by market, chain, or individual store. Sophisticated software makes very detailed market analysis possible. The disadvantage, of course, is the high price for scanning reports. Also, market coverage is not complete; for example, in 1998, Nielsen Winescan data from covered supermarkets accounted for 28 percent of total table wine shipments in the United States.

SOME MARKETING TOOLS

Some of the important tools for assessing consumer responses and behavior are surveys, focus groups, test marketing, and household panels. Surveys can be customized to cover any topic and group of consumers. They can reach consumers in person, by telephone, by mail, or via the Internet. They will address only the issues that the sponsor wants to cover. There is a risk in using surveys because they must be carefully constructed to produce valid results. Accurate surveying is both a science and an art and is very hard to do in-house. Specialist firms charge high fees for their expertise. Unlike scanner data, surveys measure what consumers say they did or will do, not their actual behavior.

Sometimes it is difficult to quantify or hypothesize about potential reaction to a wine, or what a consumer might think about it. This makes it hard to pose appropriate questions for a survey and encourages the use of a focus group to help define issues that may subsequently be tested by a survey. Focus groups consist of small numbers of consumers gathered together by a market researcher to discuss products in an open fashion. Their success in the political arena has stimulated renewed interest in them in the commercial sector.

The use of such a group allows an in-depth look at what goes into consumer opinions and decisions. The flexible format allows the questioner to follow up on interesting opinions, as is not possible in surveys. The focus group requires skilled moderators and, because of the cost, is limited in size. Consequently, group dynamics can interfere with people expressing their honest opinions. Further, it is difficult to reliably project results from a small group to the full marketplace. Like surveys, focus groups report what people say, not what they do.

Test marketing is the introduction of a product or promotion in a limited number of test markets or stores, while maintaining current practices in the remaining control markets or stores. The test units are chosen to be comparable in consumer demographics and competitive environment to the nontest units. The objective is to compare consumer response to the new practice in the test markets with consumer response to the current practice in the control markets. Private firms may do market tests in a highly controlled and precise way, or a winery may do it in a more informal way. The decision on which way to go depends on relative costs, available resources, and risks, financial or otherwise, of a wrong choice. The big advantage of test marketing is that it measures actual consumer behavior under conditions as close to "real" as possible. If properly controlled, these tests are very expensive. If the tests are not adequately controlled, the results are difficult to interpret because of the intervention of other variables.

Test marketing might be used to determine whether it is better to put wine on display or just cut the price. In this situation, the winery would identify a group of test stores where it motivated retailers to put up a display at the regular price, and another group of similar stores, matched as to size, consumer types, and other characteristics, where retailers have been motivated to reduce the shelf price. Results will guide the winery's decision about whether to extend the display concept to the rest of the country. As in all situations, the full economic cost of the alternatives should be considered, including any payments for motivating retailer behavior.

Household panel data track actual sales. Unlike surveys or focus groups, which chart people's intentions, household panels actually track individual or household purchasing patterns over time. Scanning data can record what is sold, but they cannot report to whom it was sold. Household panels record this link and do it over time so changes in behavior can be traced to particular households. For example, if scan data reported that sauvignon blanc sales were down 20 percent, household panels could determine if this drop occurred because (1) households that were buying sauvignon blanc now bought 20 percent less, or (2) 20 percent fewer households bought sauvignon blanc. The implications for sauvignon blanc sales are quite different in these two cases. The unfortunate thing about household panel data is that they are very expensive, so wineries need to have reasonably large revenues to afford them. Before buying panel data, the winery should have a very clear idea of what it is looking for.

CONSUMER PERCEPTIONS AND DISTRIBUTION CHANNELS

The channel in which the wine is sold has an effect on how the wine is marketed, how well it sells, and what image it has. This is quite different from traditional packaged goods marketing. Ivory soap, Coca-Cola, and Kellogg's Cornflakes are perceived to be the same wherever they are sold. Massive advertising, long brand histories, and relatively simple products make it possible to build enormous brand awareness and favorable images. By contrast, wine is available from a tremendous number of makers, in many different types, with a wide range of packaging. The consumer franchise is not as strong as that for the major convenience goods makers. A wine sold at a discount chain may be perceived to be of lower quality than the same wine sold in a small shop. Conversely, a wine sold at an expensive restaurant may be perceived to be of higher quality than the same wine sold in a supermarket. These consumer reactions should be measured before embarking on a distribution strategy.

A customer buying a bleach product will typically figure out how much is needed and then buy it at the cheapest place. The consumer will buy convenience products at the most convenient store without paying much attention to price. The wine consumer, on the other hand, may choose the place to buy before making the choice about what to buy. The place is important because it may offer a large selection or help in filling a particular need.

Table 2–1 provides a personal estimate of the share of recent wine sales by outlet. The estimates are based on scan data for grocery and drug outlets, proprietary surveys of on-premise outlets, and company estimates for independents and warehouses.

The attraction of chain grocery stores is in their wide distribution. If a winery succeeds in selling an item to Safeway, for example, it may find that item in several hundred stores all over Northern California. On-premise outlets are a different story. There are restaurant chains that might approve a wine-by-the-glass program for one month. The winery may suddenly be in 10 restaurants or even 30 or 40 of them. But very often, it is just one sale to one restaurant at a time. Often restaurants cannot carry large inventories and do not want to change the wine list frequently. This leads them to seek long-term supply commitments. If a winery succeeds in selling a merlot, for example, and then cannot resupply after two months, the restaurant is going to be embarrassed because it cannot serve

Table 2–1 Estimated Share of Wine Sales by Outlet

Type of Outlet	Percent of Total
Grocery stores	33
On-premise outlets (restaurants, bars, etc.)	29
Independent shops and liquor stores	20
Discount warehouses and clubs	8
State stores in liquor control states	5
Drugstores	5

a wine on its list. The winery is unlikely to get another sale in that restaurant for quite a while.

A traditional liquor store will typically have a limited number of locations. In New York, for instance, people are not allowed to own more than one liquor store; even in places where chains are allowed, liquor store chains typically will not have anywhere near the size and breadth of a chain grocery store. And there may be store-by-store purchasing even within stores with multiple locations. Wineries and their distributors face a store-by-store selling battle once again.

The promotion-oriented chain outlets tend to go with price and volume leaders when promoting wine. They want to be where they think the public is going. The people who are buying are not necessarily wine experts and are unwilling to impose their taste on the public. They typically buy the wines that scan data or other credible evidence show to be doing very well. If a winery can provide scan data that show it to have the best-selling 1.5-liter chardonnay, then the store is likely to accept that chardonnay and put it on ad or promotion because it is a sure bet.

It is difficult to gain on-premise promotion. The buyer is more likely to trust his or her own taste in selecting wines to go with the cuisine being served. The wine list is far more personal than the product list in a retail store. A winery may sell a merlot to a seafood restaurant because the restaurant needs to have a couple of reds on the list, but the restaurant is unlikely to make much effort in selling it. Of course, the attitude toward a high-value chardonnay would be different.

The floor staff members in a traditional liquor store are very important. They guide customers through the huge collection of wines and have a strong influence on the choice of wines sold in the store. The situation in the grocery store is completely different. There is a large wine inventory and a limited number of staff members. Customers are often in a hurry. That means that brand awareness and image, and the visual impact of the label are critical. The winery must reach consumers and pull them to its wines in the store through packaging and building brand image. The important lesson is that where the wine is sold is not just a question of distribution and availability but also of how the place of sale affects the perception of the wine and its sales potential. The wine shopper is a good customer for outlets like grocery stores, drugstores, and warehouse clubs because wine shoppers tend to shop more frequently, spend more money, and buy less on discount. It is good business for these outlets to promote wine because it will attract customers likely to buy other products as well and to spend more money than many other customers.

CONCLUSIONS

This chapter has been concerned with researching the consumer. It has focused on consumer attitudes and behavior toward wine. The traditional perception is that consumers progress from ordinary to extraordinary wines as they gain experience with wine and as their income grows. Research has shown that there is a considerable overlap in consumer preferences and that choices are often driven by taste rather than demographics. The path of the wine consumer is far more complex and diverse than originally thought.

Different types of consumer purchase decisions are influenced by a variety of factors. Planned purchases are strongly influenced by brand-building advertising but not so influenced by package. Display, shelf position, and POP materials influence in-store decisions. Cross-merchandising and displays influence impulse decisions. Wineries need to understand these differences in developing marketing strategies.

Sound market strategies depend on credible market intelligence. Much intelligence is based on identifying market trends made evident by shipment, depletion, or scanning data. These data differ in their validity, timeliness, and cost. The choice depends on costs, resources, and risks of a wrong decision. Data can also be collected through special activities. These include surveys, focus groups, test marketing, and panels. Each tool has advantages and disadvantages that need to be assessed.

Finally, wineries need to study how the choice of distribution outlet influences consumer buying behavior. Grocery chains are the most significant outlet and provide the least guidance to consumers. Advertising and promotion are important to these outlets. Wine shops and liquor stores are important in wine distribution and provide significant help to consumers in making wine choices. Working with wine shops and liquor stores, wineries need to know how to influence the floor staff. On-premise stores present different problems in establishing a sales relationship. Consumer and other research that details these differences will provide important information for marketing decisions.

CHAPTER 3

Market Audits

Michael R. Hagerty

INTRODUCTION

Filling the needs of customers in order to make a profit is the crux of marketing. It involves intensive analysis to determine the product attributes that consumers value and to guide product and marketing modifications to provide these attributes. It leads wineries to become as knowledgeable about their customers as they are about their own wine. This chapter presents an approach to analyzing channel members and final consumers. Channel members include distributors, restaurants, retailers, supermarkets, mail-order companies, and any other entities getting the winery's product into the hands of final consumers. Consumers can be divided into many different segments, including special-occasion buyers, regular users, those who buy wine to impress others, and those who drink wine occasionally but really prefer beer. Wineries can reach these different segments by modifying the product, the promotional programs, and the channels of distribution.

Michael R. Hagerty has been a professor of marketing at UC Berkeley and MIT. He currently teaches at the Graduate School of Management at UC Davis, specializing in how consumers and citizens perceive quality in their lives.

THE BASIC QUESTIONS

A market audit is used in developing a business and marketing plan. Following are the four most important questions in an audit:

1. What market segment is the winery targeting? Try to describe this segment in terms of its size, the benefits that consumers desire from wine, how frequently they consume it, and where they prefer to buy it.
2. What are the winery's two closest competitors as identified by customers? If they had a choice in their preferred retail outlet, what brand would they choose if they did not or could not choose the winery's brand?
3. What is the winery's unique selling proposition (USP)? What characteristics make the winery and its product sufficiently unique that customers choose it over competitors' products? The USP weaves together elements of standard marketing strategies relating to product, packaging, promotion, place, and price.
4. What system does the winery have in place for monitoring the changes in the three areas outlined above? This is an essential task because customers' preferences are dynamic, the list of perceived competitors will change as strategies change, and the effectiveness of the winery's USP will change as well.

A simple technique can be developed for asking questions that only consumers can answer. This technique can be used in surveys, informal groups, and focus groups. Basic to this technique is an assurance to the customer that it is not a sales call but an attempt to learn how to make the product better fit consumer needs. The suggested phrasing is "I want to find out how to improve the product in a way to make your life more enjoyable."

The first question seeks to identify perceived competitors. It is easily phrased as "How do you meet your needs for this product currently?" "What brands do you purchase, and where do you buy them?" "What other beverages might you buy instead of wine?"

The second question seeks to evaluate the attributes or benefits that are important to the consumer. These attributes may include taste, value, prestige, suitability for special occasions, availability, and packaging. Some of these benefits can be identified in a question such as "Why did you buy product X instead of product Y?" The response might be that product X is dry, that it goes well with food, or that friends have recommended it. First responses are often superficial. For example, questions about taste should be interpreted carefully because many buyers often prefer sweeter wines but believe that they should buy dry ones, like chardonnay, to impress their friends. Experience like this should guide the interviewer to probe for more complete answers. The ultimate objective is to determine how competitors differ on various attributes that are important to consumers in the targeted segment.

The third major question concerns unmet needs. The question might be "How can we modify our wine to make it more consistent with your needs?" This will help the winery find out what product attributes it might change to make it more attractive in the market. The first answers to this question are often vague, but after a little thinking consumers come up with good suggestions, such as more single-serving wines, better labels, or different types of bottles. Most customers do not know or care how wines are made; they are concerned about the benefits that are important to them. Typically these benefits are different from what the technology or research and development people think.

A MATRIX FOR THE MARKET AUDIT

Exhibit 3–1 shows part of a marketing matrix form used to evaluate products in a targeted market segment. It lists the various attributes against which the consumer is asked to rate the wine. A score of 5 indicates that the consumer perceives the attribute to be very satisfactory, and a score of 1, very unsatisfactory. Attributes that do not apply to the product should be left unscored. The form in Exhibit 3–1 is for only a single product. In practice it is used to rate several products viewed as substitutes and to indicate the importance of each variable in the consumer's purchase decision. Substitutes may include other beverages such as beer. A comparison of products will give the winery a better idea of what its product strengths and weaknesses are relative to its competition.

Knowledge of the importance of an attribute is essential because a consumer may give a very high rating to a particular attribute but may not think that it is very important in influencing purchase choice. The segment surveyed in Exhibit 3–1 is people aged 21 through 29 who do not drink wine regularly but do drink beer. It is a large segment and therefore an

Exhibit 3–1 Perceived Product Positioning

Product Attributes	Score 1–5 Product 1	Score 1–5 Product 2
Evaluated group: Consumers aged 21–29 who do not drink wine regularly but drink beer		
Quality		
Nose		
Taste		
Finish		
Reliability		
Alcohol content		
"Complements a meal"		
"Adds to a special occasion"		
"Good for a romantic occasion"		
"Good to relax with"		
Price		
Unit price		
Volume discount		
Coupons		
Financing		
Bundling		
Convenience		
Available in small bottles		
Available in supermarket		
Don't need to decant		
Flip-top cans		
Unbreakable bottles		
Signaling		
Varietal		
French appellation		
"Napa Valley"		
Fancy label		
Fancy bottle		
"Dry"		
Price		

important marketing target if products can provide the benefits important to members of the segment. The exhibit includes four major attribute classes: quality, price, convenience, and signaling. Signaling attributes are perceived correlates or proxies for quality.

A more complete product positioning matrix would indicate how the product is rated against the competition. Not only do customers differ in their preferences; they differ in how they perceive one winery or product as compared with its competition. The same holds true for distributors and retailers in the channel of distribution. In practice, the matrix is expanded to evaluate several products viewed as substitutes and to indicate the importance of each attribute in the consumer's purchase decision.

The first thing that people think of among quality attributes is taste. A few buyers will look beyond taste to nose and finish, but they are not likely to account for much in the beer drinking segment illustrated here. In the cola market, for example, an estimated 35 percent of consumers can distinguish between the tastes of the market leaders. This suggests that 65 percent are focusing on other attributes to make their product choice. This seems to follow in the wine market, where studies show that the information on the label has a relatively greater impact on price than does knowledge of sensory characteristics. That is bad news for producers that just want to emphasize the sensory characteristics of wine. It reinforces the need to think about all consumer preferences, not just their sensory preferences. So far, consumers appear unwilling to pay for those sensory characteristics but willing to pay for other things.

One attribute important to many American customers is reliability. They expect that the brand will always taste the same. Such buyers tend to be marginal consumers without much knowledge of wine. They want a brand that they can depend on over time to satisfy the needs that are important to them. Alcohol content may be important for some wines (for instance, very robust zinfandels where a higher level is preferred or very light rieslings where a lower content is preferred). It is unlikely to be important to the segment covered in Exhibit 3–1.

When choosing a wine, consumers may place a strong emphasis on other attributes that are more psychological. One of these attributes is "complements a meal." Consumers might rate a Mondavi wine quite high on this attribute and a Budweiser beer much lower. A similar attribute is "adds to a special occasion." This attribute has been targeted by sparkling wine producers but has not had much emphasis from table wine producers. Related to this is "good for a romantic occasion." Consumers are beginning to place more value on this attribute as wine advertisers devote more attention to it. Nevertheless, this attribute remains more strongly connected to sparkling wine. It does not seem an important attribute for beer and therefore provides one way that a wine can position itself against beer. Beer is better positioned than wine against the attribute "good to relax with." Beer producers have invested hundreds of millions of dollars showing happy people enjoying themselves in all types of situations. Wine contributes to relaxation also, but consumers, in general, have not picked up on that fact.

Price attributes may apply differently to final customers than to distributors and retailers. The matrix includes unit price, volume discounts, coupons, financing, and bundling. Volume discounts may be important within the channel of distribution. Coupons are important to supermarkets and provide a way to get point of purchase attention. They may be an important attribute in influencing supermarket and ultimate buyer decisions. Financing is not important for the final wine customer but is for distributors and retailers. The bundling attribute allows wine to be paired with complementary items (for example, cheese or bread). The winery or retailer can exploit this attribute by putting together complementary packages.

Convenience attributes relate to packaging and availability. Packaging in small bottles might be important for the intimate dinner. People often complain that a regular bottle is too large for a quiet meal for two. Availability in the supermarket tends to be very important for the segment that does not drink wine regularly and is not inclined to search it out in specialty stores. Decanting is a practice that Americans have gotten away from, and flip-top cans are probably important only to confirmed beer drinkers. Unbreakable bottles do not seem to be much of an issue. Certainly the use of plastic bottles for wine is viewed as a negative quality attribute.

Signaling attributes are often a source of confusion. They are not the same as quality attributes, which are measurable in themselves. Rather, they are attributes that are proxies for quality; they signal an expectation about quality. For example, the use of a varietal designation may suggest that a wine is of superior quality when in fact it is not. The key point is what the consumer thinks when confronted with a label describing the variety. A French appellation or a "Napa Valley" designation is often interpreted by buyers to denote a special quality of wine. In fact, such terms may be used with wines of quite different qualities. The design of labels or bottles may also signal quality attributes that are not valid. The same can be said for a wine characterized as "dry." This may be a correct description of the wine, but it may not be related to the wine's quality as perceived by the buyer. Price is a notorious signal of quality. High price may indicate a wine of very high quality, but evidence about a lack of correlation between price and quality abounds. The final test is how these signaling attributes affect a consumer's purchase behavior.

TARGETING A MARKET SEGMENT

One of the first and most crucial elements of a marketing plan is deciding on the market segment to be targeted. Suppose the winery decided to target the segment that wants a wine that is memorable for a romantic occasion. The winery must estimate the size of the segment and the volume it is likely to generate. The segment should provide a good match for the winery's output. Most important, the winery must be able to provide the benefits that the target segment desires. The suggested segment can provide good positioning for many firms. It is small enough and has unique requirements that small and moderate-size wineries can fulfill. It is not smart for a small winery to try to convert beer drinkers into wine drinkers. Beer drinkers are much too large a market.

The winery must determine how to modify its product and package to position it in consumers' minds as memorable for a romantic occasion. The winery should consider an elegant label, a thinner and smaller bottle, and advertising that establishes the connection between the wine and romance. Another option could be to provide a memento of the occasion, perhaps a peel-off label or a decorative cork or capsule. Cross-merchandising is another effective strategy. The winery might have mail-away coupons for flowers or build a display in a supermarket next to the flowers with a discount for buying both items.

Channel selection is another important element in the winery's positioning effort. A romantic occasion often involves going out for dinner. Consequently, the winery should target upscale restaurants where people go for important dates. Picnics can also be romantic occasions picked up in cross-merchandising and advertising themes.

If a winery chooses a different segment, it will need to develop a unique selling proposition that will differentiate it from competitors. For example, in targeting heavy users that are knowledgeable about wine and have a high income, the firm will need to develop a marketing plan to show how the wine fills the segment's needs better than any other wine. This might involve establishing a bed-and-breakfast facility near the winery, sponsoring musical concerts, or opening a high-class restaurant. These strategies seem appropriate to the segment and are not used by many competitors. Catering is a relatively unique activity. Not many competitors do it, and the winery will be filling the segment's need to entertain people. With all these strategies, the winery is addressing a larger set of customer needs and making life more convenient for customers.

CONCLUSIONS

Marketing experts have long been aware that a multitude of factors affect consumer purchase choices. For many products, these factors could be represented by demographic features such as age, education, and income. The story is different for products such as wine where consumers consider numerous product attributes when choosing among products. Marketers need to discover these attributes and the relative effect they have on consumer choices.

The positioning matrix is a useful tool for obtaining customer views on quality elements on these attributes. The matrix matches quality, price, convenience, and signaling attributes against the winery brand and competing brands and beverages. It may also indicate the importance that is placed on these attributes.

CHAPTER 4

Tracking Retail Sales

Kirby Moulton

INTRODUCTION

Changes in technology now permit far more effective market analysis than in the past. Computer-based scanning techniques, Internet-based household panels, and more sophisticated trading area coverage by the Census Bureau yield highly detailed data about products and customers. These data can be analyzed using advanced statistical techniques to yield the information needed to make effective marketing decisions. This chapter discusses how these changes can affect wine retailing. Most of the focus is on the use of data generated by electronic scanning systems that record retail product movement. These data have been available for more than a decade as universal product codes (UPC) have been adopted widely. Marketers can assess how they are doing against their competition, identify trends in category, product, and brand movement, compare prices and movement in different stores, and calculate market shares. The data can be analyzed in conjunction with U.S. Census and panel data that record consumer demographic information, preferences, brand recognition, and purchase behavior to guide decisions about product characteristics, positioning, pricing, and promotion. Scanning data covers sales mostly in supermarkets and other large outlets. It generally misses wine sales in specialty stores and on-sale premises. The results are biased in that they miss sales of the more expensive wines that are relatively more important in the latter outlets. Specialized firms are active in collecting and organizing scanning data and selling it to interested clients.

PAST DEVELOPMENT

The idea of electronic scanning can be traced back to the time when the first television sets and copy machines were appearing, if not before. These devices used a system in which an object was scanned and the results converted to an image on a screen or the printed page. Product scanning relies on a coded label (bar code) containing data about the product (e.g., brand, size, quality, price). An electronic scanner reads the label and transmits it for storage and combination with data scanned for other products or items. Once a system of data presentation (the UPC system) was adopted, it became practical to collect and analyze large sets of data concerning product movement out of retail stores. These codes became the basis for a tracking system using electronic scanners to collect data from products as they passed a sales point.

Initially, the data were used for inventory control and other internal management purposes. However, as use increased, it became apparent that the data could be used for more sophisticated marketing purposes. This motivated market information firms to place electronic scanners in stores, primarily grocery stores because of their wide product ranges, so that they could collect and analyze the data and sell the results to client firms. Typically, the market information firm paid cooperating retailers in money or in services. Usually, payment was made in money, although some accounts preferred to receive an organized report of their activities with the software that allowed them to analyze the data. These arrangements have persisted over time.

One use of the scanner data, for example, was to measure sales results in test markets where households received different product messages from those in the base market. This interest, however, led to the need to accumulate information on households to measure the extent to which these changes in purchase behavior were influenced by promotion or by demographic factors. This allowed market researchers to follow distinct groups of households into the store to observe their purchase behavior toward the advertised product. Such a system is valuable for the wine industry because it distinguishes between a multitude of brands, products, and promotional programs.

DATA COLLECTION METHODS

Market information firms generally collect data in two stages. This is necessary because raw data on the retail movement of a given product need to be interpreted in the context of other activities in the store that might have affected movement. This consideration makes it essential first to collect scanner data from sample stores that have agreed to provide the data and then to collect data, usually by personal visits, on factors likely to have influenced sales. The scanner data include UPC, volume, price, time of sale, and place of sale. The personal visits are to collect information on retail advertisements, displays by location, use of coupons by store and market, and temporary price drops. Dollar sales are calculated from scanner data and combined with UPC records that provide a complete product description (e.g., brand, country or region of origin, wine type, package size, attributes). From this information, a report can be produced on all the UPCs that move in categories asked for by clients of the market information firm and on the client's own sales in the covered markets. This permits detailed analysis of market shares within product categories.

The "causal" information gathered in store visits is analyzed against the scanner data to obtain estimates of how much ads, displays, coupons, and price drops influence sales. With information about the relative impact of these marketing factors on consumer loyalty, brand preferences, and product movement, marketers can decide what combination of promotion allowances, price discounts, coupons, and frequent buyer programs is likely to be most effective. Obviously, this sort of analysis is more complex and costly than market trend analysis. Household panels are more expensive to use but can help identify characteristics such as location, income, age, and household size that tend to differentiate households with favorable buying behavior from those with less favorable buying behavior. Panels also can produce data on consumer attitudes, awareness, and preferences regarding products and brands. These data are available for a fee from various firms that organize and conduct household panels. These panels vary in size and statistical reliability. In each case, the firm collects all relevant demographic data about panel members. Using small panels makes it difficult to draw extensive implications, but small panels can be more thoroughly interviewed than larger panels and valuable insights about attitudes and behavior may be gained that would be missed in the less thorough probing of larger panels. Large panels are constructed to be statistically similar to the larger population of interest to the marketer. The members may report their purchases regularly through a diary, or they may be questioned during telephone interviews or by Internet queries. The use of the Internet may make very large panel sizes feasible because the Internet can accommodate a vast number of simultaneous responses. Once a market information firm has built a customized database, then it may offer clients Internet access to the database.

The earlier systems were primarily sample based. For example, brand shares, product movement, and other sales indicators from a sample of 10 percent of a large chain's retail outlets in a given market area might be used to estimate the chain's overall results in that market. This may be adequate if the stores and the markets served are relatively homogeneous. As computer and communications techniques improved, however, it became feasible to enlarge the system to ensure a complete census of all the stores in a chain. Such a census is more expensive than a sampling procedure but provides far more data and improves the credibility of analysis. The advantage of a census is that it avoids the biases that inevitably are created by differences between the sample population of stores and the total population of stores. These differences may be reduced by more careful selection of the sample and through use of larger sample sizes. A census also improves the credibility of results for persons who are uncomfortable with analysis that projects sample results to a larger population.

AN EXAMPLE OF HOW CENSUS DATA ARE USED

The following hypothetical example illustrates the usefulness of census data. Jason Vineyards is seeking to improve its position with the large grocery chain, ABC, in the Metro City market. Data on product movement in a set of sample stores indicates that Jason's share of table wine sales in the $3 to $7 per bottle range was 14.7 percent. However, the data did not suggest where the company should concentrate its efforts to improve its position. By collecting data from all ABC stores in the market, Jason was able to identify those stores with the greatest sales potential and those with the least. The stores were compared according to two criteria. The first was each store's sales of the wine category ($3 to $7 per bottle) compared to the average sale per store for all ABC stores in Metro City. The average category sale per store was measured at 100, and the results for each store were evaluated relative to this value. The top stores had category performance scores ranging between 225 and 296. The scores in the less successful stores ranged between 37 and 46. This data allowed Jason to identify those stores that did well selling wine in the relevant category. These stores could become the target for more intensive marketing efforts.

Jason could do better by analyzing the second criterion, Jason's brand share in each of the stores. This data showed where Jason was doing well relative to the category and where it was not. By knowing this, Jason could further focus its marketing efforts toward stores that did well in the category and could do better with Jason. Jason's brand share measured across all stores averaged 14.8 percent. But in the best stores, it ranged between 20.3 and 22.0 percent. These stores performed well in the category but were not all in the top rank. Jason learned that the next rank of stores in terms of Jason brand share did relatively poorly in category performance. The stores ranked third presented an interesting mix. They did relatively well in category performance, with an average score of 230, but not so well with Jason, which gained a share of 9.5 percent, about two-thirds of the market-wide average. The lowest rank of stores had a category performance score ranging between 37 and 46 and a Jason share that ranged from 6.2 to 8.5 percent.

In comparing these ABC stores, Jason ranked them according to market share and then examined category scores. The results are listed in Table 4–1.

Jason's sales volume was greatest in the stores ranked first, but less than one-half of that in the stores ranked third, where category sales were largest. In terms of market potential, Jason reasoned that it was less likely to expand sales in stores where it was outperforming the category (Group 1) than in stores where it lagged behind category performance (Group 3). Market potential was not promising in Group 4, which did not perform well in the category or for Jason. Not much more could be done in Group 2 stores because Jason was already doing better than average and out-performing the category, and the stores did not seem to do well as wine retailers. The poor Jason results in Group 3 stores may result from poor shelf presence, lack of displays, or aggressive competitors. Whatever the reasons, these stores offer a chance for improvement in an environment where the category is doing well. A reasonable goal might be to get Jason wines just up to average brand strength. Jason has the opportunity to

Table 4–1 Jason Winery's Category Performance and Market Share in ABC Stores

Rank	Category Score	Jason Share
1	202	21.2
2	77	18.6
3	230	9.5
4	42	7.3

ask for ABC's help to make the goal attainable. It is also important to maintain performance and guard against loss in the Group 1 stores.

After reviewing the performance data developed from the store census, Jason modified its marketing strategy to ensure that it maintained its brand share in the Group 1 stores and gained share in Group 3 stores. Jason could not have arrived at this decision if it had continued to rely on sample store data and had not reached out for census data.

OTHER SALES COMPARISONS

Another use for census data is to identify stores in which a producer's brand is not sold. Such information may be hidden in sample-based data that is projected over an entire chain. The simplest way to use such information is to calculate average sales value in chain stores selling the brand, and apply this value to each of the stores that do not sell the brand. Consider a chain of 100 stores that sells a brand in 75 of its stores but not in the remaining 25. If the average brand sale in the 75 stores is $4,000, then the potential sales in the remaining stores might be $100,000, if the stores perform as the others do. This, to the producer, represents the cost of inadequate distribution. A more sophisticated analysis would use census data to measure performance in individual stores, then use weekly reports to identify factors causing differences in performance, and, finally, apply the results to estimating market potential for the remaining stores. Weekly sales reports are useful in identifying the situation in individual stores with respect to special promotion programs, the inventory situation, and the existence of other factors likely to affect sales. For example, for seasonal products like sparkling wines, it is critical to ensure that shelves are properly stocked during the selling season. Sales reports can quickly identify stores with no movement, allowing suppliers to immediately react without waiting for store personnel to take action.

Inventory turnover is a commonly used measure of performance and is useful in comparing results between stores. It is calculated by dividing the sales of a product or brand by the average inventory held over the sales period. Products with high turnover rates are attractive to retailers if the average profit margin is satisfactory. Suppliers value information on turnover rates; it is a valuable selling tool for suppliers seeking a better position with a retailer. Such data are not available through regular scanning reports, which cover transactions but not inventory levels. However, dollar velocity measures a somewhat similar performance variable and can be calculated from scanner data. Dollar velocity is the value of product sales per million dollars of store sales. It can be calculated from census data and used in comparing results between stores, regardless of store size. In the example above, the chain management may find that the brand in 75 stores has an average sales velocity of $85 per $1 million of sales, and this velocity is higher than competing brands sold in all 100 stores. This result would not be evident in a simple comparison of sales results because of the different store sizes. The chain management is now in a position to reallocate brands and drop those with lower velocities. This type of information can be very valuable to suppliers in seeking to obtain more extensive distribution, shelf space, and display features at the expense of other brands.

Dollar velocity can also be useful in assessing the relationship between consumer demographics and wine sales. A chain of 100 stores in a regional market undoubtedly has stores of different sizes located in areas of different consumer demographics. A simple comparison might show that the sales of a particular premium wine in a large store in a middle-income area were significantly larger than sales of the same product in a smaller store located in an affluent area. An inference might be that middle-income consumers are the better target for marketing programs. If, however, the differences in store sizes were corrected by using dollar velocity as the performance measure, then it would show that affluent customers were more frequent purchasers. This would confirm the positive relationship between consumer income and the purchase of premium wines. In summary, it is often more profitable to analyze the market in detail rather than rely on less costly aggregate data that may hide information that is important for marketing strategies.

USING U.S. CENSUS DATA

The census discussed up to this point has been the census of store activity. Combining it with data from the U.S. Census often can augment the value of such a census. Such data provides a description of population characteristics such as household size, income, age, education, and occupation. Much of this data is available down to postal code areas (ZIP codes). Consequently, it is possible for an information retrieval firm or other firms to identify the characteristics of consumers living in the same area as a store, or living in areas otherwise served by the store. One strategy for obtaining this type of information is to offer lotteries and other prize competitions for which customers have to register and provide address information. Another is to record the license numbers of cars parked in a store's parking lot, obtain the owners' addresses, and then classify those addresses according to ZIP code. These methods permit the store to build a demographic profile of its customers. A chain can use these methods to compare the characteristics of customers for each of its stores. Ultimately, this compares customer characteristics with sales results in different stores. With this information it becomes possible to estimate how sales are likely to respond to changes in consumer income or changes in prices. For example, the analysis might show that more affluent customers are less sensitive to price changes than moderate-income customers. Knowing this, a supplier or a store manager can estimate whether a drop in price is likely to increase or decrease revenue. Similar analyses can be used to estimate the impact of other changes in marketing strategies. It would be considerably more expensive to use panels to obtain data comparable to that provided by the U.S. Census. Panel data, of course, can support more precise analysis in a very specific area and provide information about consumer attitudes, awareness, and preferences.

CONCLUSIONS

Accurate information is the cornerstone of effective market strategies. In reviewing the long-term trends in wine marketing, it appears that decision making has moved from intuition based on little information, to decisions based on selected information available through sampling, to strategies founded on far more precise and comprehensive information. Rapid advances in scanning technologies, survey techniques, and computer capabilities have hastened this trend. Market information companies have developed to collect and sell data on a wide-scale basis. Such companies also assist in analyzing U.S. Census data and in organizing and evaluating consumer panels. Larger wine companies and suppliers have been able to conduct some of these collection and analysis functions themselves. This chapter has focused on how this wide array of data can be used to evaluate marketing situations and to devise more effective market strategies. Underlying this discussion is the implicit recognition of how fast technologies have changed and how much more precisely wine marketers can work now than they could just decades ago.

PART II

Considering the Marketplace

CHAPTER 5

The Context for Marketing Strategies: A Look at the U.S. Wine Market

Jon A. Fredrikson

INTRODUCTION

The wine industry is too complex for simple generalizations to contribute much to marketing strategies. The purpose of this chapter is to examine this complexity and derive from it useful ideas for decision making. The first part presents a historical perspective on what wine was and what it is today. The succeeding sections examine the new wine culture, the renaissance of the 1990s, and major wine market trends. The conclusion focuses on the future. Attention is focused primarily on the United States, although some of the marketing lessons are applicable to other markets. The data reported in this chapter are from the Gomberg–Fredrikson Report, company estimates, and data derived from Nielsen WineScan reports, the California Board of Equalization, the U.S. Bureau of Alcohol, Tobacco and Firearms, the Wine Institute, and other sources.

The United States is not a wine-drinking country. Although total consumption is large, and the United States is the third-ranking wine consumer in the world, it ranks very low in per capita consumption relative to many countries. Consumption in the United States is less than 15 percent of that in France, for example. Wine consumption is low relative to the

---

Jon A. Fredrikson received his MBA from Columbia University and worked for Joseph E. Seagram and Sons for 13 years, prior to acquiring Louis R. Gomberg & Associates in 1983. Since then Jon has emerged as one of the most well known consultants/economists in the wine industry. He publishes the Gomberg–Fredrikson Report, a private wine industry business report issued monthly to clients.

consumption of other beverages in the United States. Per capita wine con-sumption is just shy of 2 gallons, as compared to beer at 22.4 gallons, soft drinks at 51 gallons, and bottled water at 11 gallons. Per capita consump-tion of spirits is about double that of wine when measured by alcohol con-tent. Relatively few people account for most consumption: only 11 percent of adults aged 21 to 59 consume 88 percent of all wine. Such a small group is difficult to target through mass marketing. Reaching this group requires well-focused niche-based marketing.

A GLOBAL HISTORICAL PERSPECTIVE

Worldwide wine consumption declined by 26 percent, or 2 billion gal-lons (76 million hectoliters), between 1980 and 1993 because of major changes in consumer preferences in European countries. Per capita con-sumption rates in the major wine-producing countries declined by up to 50 percent. Although the decline abated after 1993, consumption growth in Great Britain, most northern European countries, and the United States, where per capita intake more than doubled between 1960 and 1998, was insufficient to offset that decline. Even with the declines, per capita con-sumption in France, Italy, and Spain remains far above what it is in the United States. A greater proportion of the population drinks wine in these countries, and these people drink more wine. The role that culture plays in this difference is even more evident in the fact that the number of people in the United States who drink wine at least once a year (i.e., those who know what wine is and are not opposed to drinking it) probably exceeds the total number of people who drink wine in France.

Global production trended downward also between 1980 and 1999. Estimated production in the four-year period of 1996 to 1999 was one-quarter lower (90 million hectoliters) than in 1980 alone. However, global production has exceeded consumption for many years, resulting in a sub-stantial surplus. Some of this is used commercially in brandy, concentrate, and juice production, but the major part is diverted through European Union actions to support market prices and certain rural policies.

WINE IN THE UNITED STATES

In the United States, the federal government defines wine much more broadly than governments of most European producing countries. The definition covers beverages fermented from fruit (including grapes) that may include other flavoring and carbonation and is subject to certain alcohol limits. This covers a conglomeration of beverages, including slightly spritzy "pop" wines, peach wines, vermouth, fortified wines, and the most sophisticated ultrapremium wines from exclusive production areas. Table wines are defined as still natural wines with alcohol content less than 14 percent, and dessert wines are still natural wines with alcohol content over 14 percent. It was not until 1967 that table wine consumption actually surpassed so-called dessert wine consumption.

After World War II, wine in the United States was largely a blue-collar beverage; it was a California port, or a New York sherry, or an indiscriminate muscatel that provided an inexpensive form of alcohol for 99 cents a quart or less. It was a much cheaper form of alcohol than spirits because excise taxes on wine were considerably lower than those on spirits. Wine evolved into the modern-day beverage in the United States only during the 1970s.

Because wine demand was growing at about 11 percent per year through the end of the 1960s and into the 1970s, vineyards and wineries expanded their capacities rapidly. Industry professionals and outside experts were projecting demand based on current growth rates to reach 9 to 12 gallons (34 to 45 liters) per capita by 1990. (By 1999, per capita consumption had not yet reached 2 gallons.) The industry geared up for an enormous increase in demand and planted thousands of acres, built huge wineries, and attracted major corporate interests. Coca-Cola entered the industry through the acquisition of Taylor Wine Company of New York and the creation of Taylor California Cellars. Coca-Cola Company included a wine importing company in their portfolio and reached shipment levels of 11 million cases. Other major players at one time or another included Seagram, National Distillers, Pillsbury, and RJR Nabisco. In addition, of course, were the large family wineries such as Gallo, Sebastiani, and Franzia.

Then the industry hit the skids in the early 1980s. People reevaluated what was learned from the experience in the 1970s in order to understand what was happening in the 1980s. Why did wine consumption decline in the 1980s, especially in the face of projections of wine nirvana? A major cause was a change in governmental attitudes. The national drinking age effectively was raised to 21. MADD (Mothers Against Drunk Driving) and similar antialcohol pressure groups were created in the early 1980s. A major producer of spirits and wine started an "equivalency" campaign to convince regulators and tax authorities that the source of alcohol was irrelevant and that, for taxing purposes, a drink is a drink is a drink. This was disastrous for wine. Prior to that, wine was thought of as apart from spirits and beer. It was perceived as less likely to lead to alcohol abuse and its consequences. The equivalency campaign convinced many legislators that all alcohol is the same, that it leads to the same behavior, and that it ought to be taxed in the same way. However, the changes did not end there.

By the mid-1980s many people grouped wine with recreational drugs, and so the fight became against drugs and alcohol. An educational brochure might well have a glass of wine with an X through it. There were anecdotal stories about hearing a child cry, "My dad's a drug dealer; he sells chardonnay." In addition, government-mandated warning labels were telling women, in effect, "don't even touch wine while you are pregnant," and this tended to inhibit consumption, even among those who were not pregnant.

Wine coolers were another important phenomenon in the U.S. market in the 1980s. They appeared in the early part of the decade and peaked about 1987, with about a 20 percent share of the U.S. wine market, and then declined more rapidly than table wines. The decline in coolers exacerbated the general malaise in the market. It was hurried along by a change in government excise taxes that made it more profitable to use a malt beverage base than a wine base for coolers. Wine coolers virtually disappeared, but the industry still produced millions of cases of "malt" coolers.

As soon as consumption dropped in the 1980s, prices plummeted as well, especially for the inexpensive economy wines sold in large sizes. These were the wines in the so-called "jug wine" category, produced mostly in the Central Valley; with traditional generic names such as chablis, burgundy, and rosé; and sold in three-, four-, or five-liter containers. These wines account for somewhat less than one-half of table wine consumption now, but in the 1970s they dominated the market. When the bottom fell out in the 1980s, these wines became so inexpensive that there was no money left to spend on advertising. Table wine advertising expenditures dropped from almost $200 million (measured in constant 1987 dollars) in 1982 to about $40 million (measured in the same 1987 dollars) in 1994. Table wine shipments declined from about 400 million gallons (15.2 million hectoliters) in 1983 to approximately 300 million gallons (11.4 million hectoliters) in 1991. The correlation between changes in advertising and changes in shipments was not perfect, but the loss of advertising support obviously hastened the decline in sales.

The result was that competition centered on price, and the low margins drove the big corporate entities to sell out. Coca-Cola, Seagram, National Distillers, and RJR Nabisco exited (or refocused) because they could not make money in a market with such overcapacity. So it was an antialcohol culture that emerged, and wine was out.

THE RENAISSANCE OF THE 1990s

In the 1990s, cultural attitudes toward wine changed because of a series of issues related to health and dietary guidelines. The original "French paradox" story, featured on *60 Minutes*, a CBS TV network news show, in November 1991 may have triggered the change. It pointed out the relationship between drinking, diet, and heart disease. It reported that the French had lower rates of heart disease and consumed far more wine than consumers in countries with higher rates of heart disease. The relationship between wine drinking and health was revisited again by *60 Minutes* in 1995, after an avalanche of research publications confirmed the health effects of moderate wine consumption by persons not at risk from alcohol. Almost every week, there was some new study showing the health effects of wine. The accumulation of this health news and the related attention given to wine stimulated a resurgence in the wine market. The link between wine and drugs was weakened by new medical evidence, which made consumption more acceptable.

The U.S. Dietary Guidelines moved from a recommendation to abstain from alcohol to a position more supportive of wine with meals. Wine could now be enjoyed without guilt. Lifestyle preferences also changed in a way that made wine a part of better living. Consumers were increasingly moving up to premium products among categories such as coffee, fresh juice, beer, and gourmet food. A growing economy helped these preferences be realized. This added up to a change in cultural attitudes about alcohol and especially wine. Now wine is viewed as part of a healthy lifestyle and is in step with consumer attitudes about better living and less austerity.

MAJOR WINE MARKET TRENDS

The most dramatic trend in the U.S. market has been the rapid growth in premium wine sales, measured both in volume and in value. California wineries' revenues from premium wines exploded from about $200 million in 1980 to $4.4 billion in 1999, a staggering growth that stimulated the planting of new vineyards, the opening of new wineries, and the expansion of older wineries. To understand this change, it is first necessary to understand how premium wines are defined.

The U.S. government does not define standards for wine quality. Therefore, the industry and trade have had to develop a framework for classifying quality wines. Some may believe that all wines labeled by varietal type are quality wines, as compared to those that use generic names. Others may believe that the region of production can identify quality wines. The problem is that there are too many examples of wines that would not meet the consumer's perception of what is quality. Data collection is difficult for varietal wines but less so for regional wines. In any event, the most common measure of wine quality for statistical purposes in the U.S. market is the retail price. A commonly used classification for table wines defines ordinary (or "jug") table wines as those wines selling at a retail equivalent of less than $3 for a 750-milliliter bottle; popular premium wines sell for $3 to $7; superpremium table wines $7 to $14; ultra-premium wines $14 to $25; and luxury table wines more than $25.

The high-end premium wine market (i.e., more than $7 a bottle) drove the revenue increase. It accounted for 52 percent of total winery revenue in 1999 while shipping only 23 percent of the volume. The popular premium segment ($3 to $7) accounted for 31 percent of the revenue and 33 percent of the shipments. The disparity between revenue share and shipment share is evident in the jug wine segment (below $3), which generated 17 percent of the revenue and 44 percent of the shipments.

Packaging trends are revealed in supermarket data. There has been a decline in three- and four-liter containers and growth in the five-liter bag in the box, which is now the most important big size. It is a popular size because it fits in the refrigerator against the wall and has a remarkable shelf life. It is economical, at retail prices that are equivalent to 25 cents per four-ounce glass. The 1.5-liter bottle is a growing factor in varietal wine sales.

Another important trend has been the expansion in California exports. In 1999, shipments from California wineries to export markets were 24.3 million cases (2.2 million hectoliters), more than five times the 1990 level of 4.8 million cases (0.4 million hectoliters). The major export destinations are the United Kingdom, Canada, the Netherlands, and Japan. Many wineries are hedging their bets, in recognition of increased future production, by staking out positions in these markets around the world. Even though some wineries may not have enough wine to support a wine list in New Jersey, they still believe that it is more important to develop and protect a position in growing foreign markets.

The most important trend of the 1990s has been the incredible growth of red wine consumption. It has more than doubled since that original "French paradox" broadcast in 1991. This led to a remarkable problem: trying to find more red wine somewhere in the world to continue to feed this strong growth. The growth in consumption has been driven by the wealth of health news, and it is reflected throughout the market. For example, scanner data indicate that before the "French paradox," red wines were only 16 percent of consumer purchases in supermarkets; by 1999, they were 35 percent, more than doubling their market share. This has been at the expense of both white and blush wines, which in 1999 accounted for 39 percent and 26 percent of supermarket sales. Ironically, California was pulling out red grapes until 1987 because their supply exceeded market demand in the 1980s.

Two other major trends affect the U.S. market today. One is the projected growth in wine production capacity, and the other is the growing practice of U.S. wineries purchasing wines from foreign countries for bottling in the United States under the importing winery's label. For several years it appeared that the wine-grape-growing area in the United States was under-reported. In California, for example, the reported area was 329,000 bearing acres in 1997, a figure subsequently corrected to 374,000 acres, an increase of 45,000 acres (14 percent). Bearing acreage for 1999 was subsequently estimated at 410,000 acres. If the *newly planted* acreage reported in 1997 comes into production as planned, then the bearing area in 2000 will exceed 440,000 acres, an increase of 18 percent in production potential from 1997. This could lead to an additional 63 to 76 million gallons (2.7 million hectoliters, on average) of primarily quality wine to be marketed, depending on acreage removal and vineyard productivity assumptions. This is a large quantity in relation to California premium table wine shipments of 202 million gallons (7.6 million hectoliters) in 1999.

The U.S. wine renaissance of the 1990s created a remarkable turn-around opportunity for imported wine marketers. Widespread U.S. wine supply shortages, particularly for red wines in the popular premium price range ($3 to $7), gave importers the chance to reclaim a significant portion of the U.S. market share they had captured in the early 1980s.

At their peak in 1984, imported wines totaled approximately 142 million gallons (5.4 million hectoliters). Then real cost increases, the weakening of the U.S. dollar, some wine contamination problems, and a booming market for California "fighting varietals" caused imports to plummet over the next seven years to about 60 million gallons (2.3 million hectoliters). By 1997, imported wines climbed back to 118 million gallons (4.5 million hectoliters), only to fall back to 108 million gallons (4.1 million hectoliters) after the abundant 1997 California harvest replenished wine inventories.

Imported bulk wines played an important role beginning in about 1996, when U.S. producers used 7.5 million gallons to augment their domestic wines. In 1997, 20.3 million gallons were imported to help U.S. marketers temporarily get over the California wine shortages. In spite of the giant 1997 harvest, some wineries continued to import foreign bulk table wines in the following year. About 8 million gallons were imported in 1998 and slightly less than 5 million in 1999.

CONCLUSIONS

California premium wine sales should continue to grow as added domestic production comes on line. This will reduce the need for imported bulk wines, although they will remain as a potential source if U.S. grape prices become too high relative to prices for grapes in other countries. Bottled imports will continue to grow because they have a foothold and have established some major brands.

As more wine comes to market, there will be increases in consumer advertising. It is reasonable to expect levels in the range of $100 million or more in 1999 and 2000, a level that appears consistent with the growth in wine shipments through August 2000. One of the risks is that if the spirits industry pushes the government to allow spirits advertising on television, as permitted for wine and beer, the government could react by banning all alcoholic beverage advertising on television. The outlook for high-end premium wines (priced above $7 per 750-milliliter at retail) is particularly good. There could be some marketing problems without advertising if it is banned, but this could be offset by continued publicity about the health effects of wine, which seem especially relevant to an aging population.

There will be a strong growth in varietals such as chardonnay, merlot, and cabernet sauvignon bottled in 750-milliliter, 1.5-liter, and 5-liter sizes. Chardonnay remained the most popular varietal in 1999 supermarket sales. Sales growth for merlot in both volume and percentage exceeded that for chardonnay and pushed merlot above cabernet sauvignon in market share. This trend will continue to accelerate as grape and wine production increases beyond 2000.

CHAPTER 6

The Production/ Marketing Interface

Tom Eddy

INTRODUCTION

Each production technique has a motivation, whether it is related to marketing or efficiency. Marketing decisions affect production choices about varieties, blending, fermentation, and aging. Production decisions affect marketing choices about positioning, product differentiation, pricing, and promotion. Changes in production techniques are critical to presenting new and varying styles of wine to discriminating customers and making money while doing so. The purpose of this chapter is to review some of the production choices facing modern wine makers and deduce how they affect marketing decisions, and vice versa. The chapter works from the premise that quality is a moving target, and therefore, marketing and production decisions need to adjust continually. In the final analysis, wine makers must give consumers a product that they want to buy if the winery is to be profitable in a very competitive international market.

PRODUCTION MANAGEMENT

Table 6–1 presents an outline of major trends in production practices. It is designed to provide marketing specialists with an idea of where product

Tom Eddy received his fermentation science degree from UC Davis in 1974 and held progressively more responsible wine making positions with Inglenook, Souverain, and Christian Brothers. For the past decade he has been the principal in Thomas G. Eddy and Associates, a consulting firm, as well as owner and wine maker for the Tom Eddy Winery, specializing in hillside cabernet from the Napa Valley.

Table 6–1 Major Production Changes for Wine and Grapes

Past Practices	Recent Trends
Grape/wine processing	
Screw-type presses	Bladder/membrane presses (batch style)
Wood/concrete fermenters	Stainless steel rotary fermenters
Simple destemmers	Destemmers with variable speed and adjustable rolls
Blending and distillation of bad wines	New methods to remove ethanol and volatile acids
Wood tanks and used whiskey barrels	New barrel technology and style: French vs. American oak; toasting; aging, shaving, bending, etc.
Common yeast strains	Nonintervention wine making (alcoholic fermentation)—natural flora
Barrels for flavor	Barrel substitutes: Innerstave, chips, extracts
Minimum harvest criteria	Must adjustment with organic acid; improved sampling in field; pH criteria, ammonia checks
Minor microevaluation	Lab control for flavor: for instance, volatile flavor components and Brettanomyces (4-ethylphenol)
Quality control through "cleaning" wines	Minimalism: less is more (handling)
Early racking	Surlies and lees stirring for flavor
Control all malolactic fermentation	Use of malolactic bacteria fermentation for flavor enhancement (vinaflora)
Skin contact: quantified	Whole berry: cold soak, whole cluster pressing, carbonic maceration
In-house lab and bottling	Mobile bottling lines and other services
Viticulture	
Minimal rootstock selection	Rootstock selection: soil conditions, pests, yields, flavor development
Miscellaneous cuttings used for vines	Clonal selection for best results; nursery selection
Vineyard layout for equipment ease	Trellis/planting design for optimum balance between quality and yield

Sugar as only quality parameter
Simple long-term contracts

General soil criteria
Traditional variety selections
Traditional growing areas
Traditional harvest methods, mostly hand

Marketing and business
Generic bottle molds
Stay-at-home wine makers
Traditional packaging
Traditional wine making
Estate production only
Limited supply of wine making materials

Harvest parameters expanded: acid, pH, balance, temperature, yield
Gross per acre rather than cost per ton; crop limits, responsibility for picking dates; exotic formulas
Terroir management: correlation with wine quality
Meritage blending and varietal selection
New and more refined viticultural appellations
Improved modern machine harvesters; other mechanical harvesting aids

Designer molds for each producer
"Flying wine makers" on a global basis
Creative labeling and packages
Global wine making: better communication
Custom crush for others; alternative premises
Vertical integration of wine making materials and supplies

changes originate. It illuminates two important facts: (1) there are a multitude of points where intervention will change the nature of the product, and (2) some changes are technology driven and must be adopted to remain economically efficient and keep up with the competition. The list is not exhaustive but includes those changes that have resulted in more palatable and profitable wines. As is true for all generalized lists, some practices are not appropriate to all wines but are developed for a specific type or style of wine.

The quality of wine is determined mostly in the vineyard. Choices made at the processing level can improve that quality or can destroy it. In Table 6–2, I detail what I believe to be the relative importance of various factors influencing wine quality.

GRAPE AND WINE PROCESSING

One of the major changes that has occurred in the wine industry is the style of presses that are used. For years wineries in California used screw or horizontal presses that were very efficient. They could recover 190–195 gallons per ton (8 hectoliters per metric ton) in some cases, especially in the Central Valley. This was acceptable when the industry was more focused on productive efficiency than on wine quality. The wine makers knew that if part of the juice was a little bit bitter from excess pressure, they could always distill it or put it into the dessert wine program. So everybody was happy, especially the comptroller, because the winery recovered everything it could out of the grape.

Table 6–2 Factors Influencing Wine Quality
(Percent of Relative Importance)

Factor	Importance (Percentage)
Grapes	75
Barrels	10
Wine making method	6
Wine maker	4
Winery equipment	3
Bottling plan	2
Total	100

As consumers began to opt for quality wines instead of dessert wines, the old pressing practices stopped being profitable. About this time, California wineries began installing bladder or membrane presses, which were used extensively in Europe. These presses gently squeezed the grapes and pumice to the side of the tank and brought to more acceptable levels the tannins, coarseness, and bitterness that were a byproduct of the screw press. Surprisingly, the recovery is very close to that of the screw press, but the product is better. Even though a bladder press processes 15 tons over a two-hour cycle and costs approximately $100,000 to $125,000 to install, it still pays for itself in two years.

Another area of change is in the type of tanks used. Wine makers in many countries moved away from concrete and wood tanks to stainless steel in the 1960s and 1970s. They then began installing more complex tanks, for example, with jackets that allow better temperature control of fermentation and, more recently, rotary fermenters that eliminate the need for punching down the cap and pumping over. The rotary fermenter saves on labor costs and attains better extraction, but its use is limited to large facilities that can afford the cost. Other equipment allows better quality control in destemming and in removing volatile acids from wines.

I believe that barrels have the greatest influence on wine quality of any cellar practice and deserve more discussion than other practices. The use of barrels over time has evolved in a way to enhance wine quality. This has resulted from new barrel technology and styles, including alternate sources for oak, toasting practices, aging of the oak, and shaving staves. It has also fostered the development of less costly barrel derivatives that impart similar flavor characteristics to lower-priced wines. These derivatives are primarily oak chips and extracts.

Arguably, French oak barrels set the standard for wine making, although many wine makers are beginning to find other sources for oak that provide the wine characteristics that they seek. In California, a recent price for a French oak barrel was $650 from a boutique producer, providing the right toast level and including toasted heads, extra hoops, and maybe a wooden hoop or so. State-of-the-art American oak barrels manufactured in California by French barrel makers cost about $330 each. A new American oak barrel made in Kentucky was priced at around $165. The cost of using French oak barrels is about $1 per bottle, while the cost for American oak is about 50 cents per bottle. The choice between the two depends on marketing decisions about pricing, target markets, and consumer preferences.

Another product that works reasonably well is French or American oak cubes, about the size of sugar cubes, contained in a perforated plastic tube, or staven, that is inserted into an aging barrel. The cubes are toasted to help break the cellulose structure of the wood and give the wine a desirable flavor. This system allows a winery to obtain some of the flavor characteristics of the more expensive aging systems while continuing to use older wood barrels that no longer impart flavor. The wood costs from $4 to $6 a pound and probably represents a per bottle cost of 3 cents to 8 cents.

The next level down is wood chips and extracts. Oak chips sell for as low as $2 a pound and provide some of the flavor components and little of the complexity expected in a high-priced wine. For example, I work with an inexpensive chardonnay that sells for $70 a case. At this price point, one can give consumers the oak flavor they look for by using about 25 pounds of oak chips per 1,000 gallons of wine. The cost is about 1 cent per bottle, as compared to 50 cents to $1 per bottle for higher-priced wines using barrels for flavor extraction and complexity.

Twenty years ago in California, there were basically two strains of yeast: champagne for red wines and montrachet for white wines. Both of these yeasts sometimes produced hydrogen sulfide (H_2S), a very undesirable characteristic for wine. Now there are yeast strains better suited for flavor extraction and strains that do not produce H_2S or that speed up fermentation. One strain ferments at a lower alcohol conversion rate. This reduces the production of high-alcohol wines subject to greater taxation, or the cost of lowering alcohol content to desired levels. The increased availability of yeast strains means that there may be a different yeast specification for each vineyard block, depending on differences in the juices that they produce. The end result is a better wine to help achieve marketing objectives.

VITICULTURE

Developments in improved rootstock have increased the wine quality choices available to wine makers. Historically, the wine maker was more concerned with variety type, production, and harvest time than with rootstock and how it related to vineyard conditions. Now wine makers are concerned with the rootstock, its match with the varietal clone, and its productivity. Clonal selection is taken far more seriously, with specific attention to clones that provide desired quality attributes in the grapes. The challenge is to balance quality considerations with productivity and profit.

Rootstocks are selected for their resistance to pests and diseases, their suitability for specific soil conditions, and their productivity and match with the grafted variety.

Because grapes are the primary determinant of wine quality, it is worthwhile to examine trends that tend to improve grape quality. Important among these are changes in harvest criteria and improved field sampling techniques. Historically, California contracts between wineries and growers have specified one quality characteristic, sugar. Growers received price premiums for grapes with higher than specified sugar and were penalized for low sugar. This resulted in delayed harvests and high-alcohol wines that tended to lack balance. These are not characteristics favored by today's consumers.

The industry has progressed significantly from that point and specifies standards for sugar, acid, pH in the resulting juice, and temperatures at harvest. The winery may also specify maximum yield and when harvesting is to occur. The harvest temperature specification is interesting. If cabernet sauvignon grapes are harvested on a cool morning when temperatures are under 50°F they are too cool for effective flavor extraction. The grapes have to warm up to 70°F before fermentation starts, and this may hold up a tank for two or three days. An alternative is to delay picking until the ambient temperature is 60°F or more so that fermentation will start relatively promptly. By starting at a warmer temperature, the wine maker gains faster extraction and more flavor, and lowers costs.

Efforts to reduce yield to improve quality have an economic cost. For example, if a high-quality pinot noir vineyard is capable of producing four tons per acre and selling at $2,000 per ton, its revenue will be $8,000 per acre. If the buyer wants to reduce yields to three tons in an attempt to augment quality, then the buyer must be prepared to increase the grape price to $2,700 per ton and push up the retail bottle price by $5. This may or may not be easy depending on previous product positioning decisions. Of course, as wineries are able to command higher prices, growers demand higher prices as well, while at the same time planting added acreage that ultimately will lower prices again.

MARKETING AND BUSINESS

The business approach to wine making has changed considerably. Rather than accept the designs that glassmakers push, wineries are able to select glass molds designed for individual users. There are limits to this range, created by container size regulations, but the result can be embossed symbols, flange tops, or variations of standard shapes. These options provide additional means for distinguishing one product from another. Similar trends in packaging and labeling are discussed in other chapters in this book.

Wine makers have become global. They move from one continent or hemisphere to another and spread new wine-making techniques quickly. This ease of movement and communication has been greatly aided by the emergence of e-mail and the Internet. One result of this is that competitive advantages are not as sustainable as they once were. This heightened competition encourages even further changes in marketing and technology. Wine makers are used extensively in marketing and promotion. A "star" wine maker attracts considerable media attention to the winery and becomes an important element in fixing its image. Many wineries have also broken the traditional mold by taking on custom production for others, by developing outlets for wines that do not meet winery requirements, and by using alternative premises. The ultimate in this, of course, is Virtual Vineyards, described in Chapter 9, which owns no production facilities at all.

Organizational strategies have also emerged to gain control of supplies and wine-making materials. These include the acquisitions of vineyards by wineries in California (such as Gallo, Kendall-Jackson, and Robert Mondavi) and the integration of bottle production activities and distribution agencies.

WINERY COSTS

A review of winery cost structures can provide an idea of the sensitivity of total costs and derived prices to changes in individual input costs and overhead expenses. The following cost estimates (Table 6–3) are derived from actual winery experience in 1996 and 1997 and are reasonably close to industry standards. They are for a large winery (more than 250,000 cases) producing wine to sell at $7 per bottle, a medium-sized winery (100,000 cases) producing wine to sell at $12 per bottle, and a small winery (10,000 cases) producing wine to sell at $30 per bottle. The importance of these estimates is in their relationship to one another rather than in their absolute level.

The cost for grapes ranges from 40 percent to 45 percent of total production costs for each of the listed wineries. A 10 percent change in grape costs, not unusual for an agricultural commodity like grapes, will change production costs by 4 percent to 4.5 percent. From a marketing perspective, the question then becomes one of price: should it be raised or lowered as costs change? Or conversely, if the marketing director wants to position a new and prestigious variety or blend, will the cost of grapes push the price beyond a market-acceptable level? In the example, the small winery producing a bottle of wine to sell at $30 is paying about $2,600 per ton for grapes, the medium-sized winery selling wine for $12 per bottle is paying about $1,600 per ton for grapes, and the large winery producing wine to sell at $7 per bottle is paying an average of $900 per ton for grapes. These cost-price relationships set the parameters within which the marketing director must operate.

Barrel costs, on the other hand, are significant for the small and medium-sized wineries but not for the large winery selling wine at $7 per bottle. Presumably, the large winery is using chips, blocks, or extract for flavor. If the sales force wanted a more "genuine" oak flavor for the wine, could it compete profitably if the price were raised by 50 cents or $1? The cellar process is a big item that includes all cellar and processing labor used for wine making and related activities. It is relatively more important in the cost structure of the large winery and undoubtedly motivates attempts to reduce labor. The costs of bottles are different for the three listed wineries, indicating that packaging must be appropriate for the targeted market. In each case, however, the cost is approximately 10 percent of production costs. Overhead costs include general, administrative, marketing, and sales costs. They are relatively high in the small winery because they are spread over a limited output and reflect the higher costs associated with fine wines. They drop markedly for small wineries producing more moderately priced wines, as they do for the listed medium-sized and large wineries.

The margins before tax are inversely related to the size of the listed wineries. However, data that pair a medium-sized winery with a small winery producing wine to sell at $12 per bottle, and a similar pair selling wine for $30, show that the medium-sized wineries earn a greater margin

Table 6–3 Estimated Production Costs per Bottle for Selected California Wineries, 1996 and 1997

	Winery 1	Winery 2	Winery 3
Annual case output	10,000	100,000	>250,000
Target retail shelf price (per 750-milliliter bottle)	$30	$12	$7
Production items			
Sampling analysis	0.011	0.003	0.005
Grapes	3.472	2.083	1.111
Hauling	0.069	0.000	0.000
Crushing	0.486	0.347	0.208
Fermentation additives	0.050	0.001	0.001
Yeast	0.010	0.004	0.001
Barrels	1.042	0.521	0.006
Aging loss	0.208	0.104	0.044
Aging analysis	0.029	0.007	0.015
Consultants	0.292	0.017	0.003
Cellar process	0.847	0.508	0.450
Glass (bottles)	0.833	0.375	0.313
Corks	0.240	0.140	0.085
Capsule	0.100	0.045	0.042
Label	0.075	0.040	0.035
Box	0.000	0.002	0.001
Bottling	0.175	0.138	0.121
Shipment to warehouse	0.010	0.010	0.010
Excise tax	0.074	0.274	0.274
Lab work	0.009	0.029	0.025
Production costs per bottle	8.032	4.648	2.750
Overhead	7.260	1.730	0.860
Margin before tax	6.708	1.122	0.140
Average price at the winery	22.000	7.500	3.750

because of economies of scale for some production items. Thus, small wineries are hard-pressed to remain competitive against medium-sized wineries. This reinforces the need for product differentiation and careful positioning. The large winery in the example is under pressure to improve its performance, although it is impossible to fully evaluate the margin without data on investment.

A LOOK TO THE FUTURE

The following are my production trend predictions for the year 2020. I recognize that some of them might already have happened by the time the list is published, but this is how the situation looked to me in 1997.

- All major winery estate properties are classified by the media for first, second, and third growths, as in France, for quality-defining purposes. Although initially disputed by a majority of vineyards, ultimately the new Council of Viticulture and Enology Quality Assessment (VEQA) agrees to a three-year tasting study to finalize all rankings. The system is not legal but is accepted by the trade.
- Sixty-five percent of all table wine consumed in the United States is red.
- The largest six wineries in the United States own overseas properties and produce "international blends" for U.S. consumption.
- Several wineries offer different proprietary blends made from exactly the same lot of wine but with organoleptic characteristics that have been altered by fractional blending of ion-exchanged portions. Initially rejected by the Bureau of Alcohol, Tobacco, and Firearms, the concept is finally approved by all agencies when it is demonstrated that the varietal flavor loss is minimal and that the blends are completely different and complex.
- Most wine makers use American oak for whites and most reds. French oak is used only for heavy ultrapremium wines. This assumes that American oak retains its significant price advantage.
- Most grape contracts written in the North Coast of California include certified soil profiles and soil mineral analysis and verification of rootstocks and clones as part of the agreement. Prices are paid on a per acre basis, and many wine makers provide their own employees for thinning and pruning as part of the contract.
- Several North Coast wineries use recycled glass exclusively. Collection facilities and washing plants are set up in 12 locations in the San Francisco Bay Area.
- Gewürztraminer, pinot blanc, and sangiovese grapes disappear. The new "hot" wines are from Eastern European estates. The green Hungarian grape becomes the base wine for a new wine cooler consumed by the Y generation.

CONCLUSIONS

The wine sector is extremely complex and offers examples of many different successes and failures. The preponderance of experience, however, emphasizes that marketing is really a driving force behind production. Work on the production plan can start once the marketing plan has been developed. There's a misconception that if a winery makes a great wine, profits will roll in automatically until the owner decides to retire in luxury. This does not happen; the market is extremely competitive and offers large quantities of quality wines. Wineries must first determine their marketing and financial objectives, and the resources they have available to achieve those objectives, then talk about wine making. Wineries that have understood this have succeeded reasonably well, but the others have not.

CHAPTER 7

Wine Marketing and Its Legal Environment

J. Daniel Davis

INTRODUCTION

This chapter is designed to give winery marketing managers an appreciation of aspects of the U.S. legal environment that shape most of their marketing decisions. It focuses primarily on tied-house laws that seek to keep producers from integrating forward into the distribution sector; these laws are so important in guiding decisions about merchandising, promotion, and distribution. The chapter is not a review of all laws and regulations, nor does it present legal opinions. Its objective is to help marketers recognize the importance of law to their marketing strategies.

The legal environment within which wine marketing takes place is essentially a legacy from Prohibition. There are a number of laws that keep a winery from promoting its product with retailers in the way virtually any other business can promote its product. A business license is a key feature of this environment. All wineries, distributors, importers, retailers, and others involved with the production and distribution of alcoholic beverages must have a license. A license is permission given by the government to do something that is illegal to do without the license. It is granted on the promise that the license holder will obey the law. If the law is

J. Daniel Davis practices in the Wine, Beer & Spirits Law Group at Pillsbury Winthrop LLP (formerly Pillsbury Madison & Sutro LLP) in San Francisco. He has reviewed and defended wine promotions since 1984 and was formerly in-house counsel for a major New York City alcoholic beverage importer.

broken, the license can be taken away. This is the leverage that ensures compliance within the legal environment.

There are two primary sets of laws that govern the wine business in the United States. First, there is federal law, which is expressed principally in the Federal Alcohol Administration Act administered by the Bureau of Alcohol, Tobacco, and Firearms. Second, there are state and local laws. The constitutional amendment that repealed Prohibition allowed each state to regulate alcoholic beverages within that state. Consequently, the basic federal law applies to all the states, but the 50 states, certain cities and counties, and the District of Colombia also have laws governing alcoholic beverage trade practices. Generally, if one set of laws, state or federal, is stricter, the stricter law applies.

The laws that affect marketing mostly center on the three-tier structure of distribution. The first tier is the winery or importer. The second tier is the wholesaler or distributor that purchases from the winery or importer and sells to the retailer. The third tier is the retailer that purchases from the wholesaler and sells to the consumer. A retailer is any entity selling wine to the consumer.

TIED-HOUSE LAWS

A big fear underlying U.S. alcoholic beverage law is that large producers could buy up small retailers and try to force consumers to buy alcohol. In doing this, the producers would force independent businesses to fail. Much of the regulation facing wine marketers stems from laws trying to prevent this consolidation and protect the three-tier system from destruction. These laws are termed "tied-house laws" because they limit the relationship between suppliers and wholesalers on the one hand, and retailers on the other. They seek to prevent a tie between levels in the distribution chain, which would endanger the chain. These laws control vertical integration in the alcoholic beverage sector.

Tied-house law regulates what interests a supplier can own in a retailer, including inventories, leases, and mortgages. The basic concept is that suppliers cannot own retailers. The law defines those services or things of value that a supplier can furnish a retailer, including equipment, signs, and samples. The objective is to limit the influence that a supplier can have over a retailer. It also regulates what services a supplier can buy from a retailer, such as advertising, display, and distribution services. Primarily, the winery cannot pay the retailer to advertise, distribute, or display its product. Federal and state tied-house laws generally have the same regulatory framework. The law prohibits a broad range of conduct between suppliers and retailers and then lists specific permitted activities that are exempt from the general prohibitions.

GENERAL PROHIBITIONS

A violation of the federal tied-house law requires three elements: inducement, effect on interstate commerce, and exclusion of others. The California tied-house law requires only the prohibited inducement. There is no interstate/foreign commerce or exclusion requirement to prove violation in California.

Under the federal law, the first element is the existence of an inducement. An inducement is a prohibited act or act that is questionable, such as owning a retailer, giving the retailer something of value, or paying the retailer for some activity. That is the inducement, but it is insufficient under federal law to constitute violation of tied-house law. The second element is that the alleged violation must have a relationship to interstate commerce. The actual parameters of that requirement are a little vague, but most acts can be tied to interstate commerce. The third element concerns exclusion. That means that the gift or payment to the retailer must be sufficient to induce the retailer to exclude the purchases of other products. Current federal regulations define exclusion as a practice that places (or has the potential to place) retailer independence at risk by means of a tie or link between the industry member and retailer, or by any other means of industry member control over the retailer, and that results in the retailer purchasing less than it would have of a competitor's product. The exclusion element is very difficult to prove because it must be shown that the retailer has lost its independence.

There are several criteria that indicate exclusion. One is that the practice restricts or hampers the free economic choice of a retailer to decide which products to purchase, or the quantity in which to purchase them, for sale to consumers. A second criterion is that the supplier obligates the retailer to participate in a promotion to obtain the supplier's product. Another indication is that the retailer has a continuing obligation to purchase or otherwise promote the supplier's product. A fourth criterion is that the retailer has a commitment not to terminate its purchasing relationship with the supplier. The fifth criterion is that the practice involves the supplier in the day-to-day operations of the retailer. For example, the supplier controls the retailer's decisions on brands of products to purchase, the pricing of products, or the manner in which the products will be displayed on the retailer's premises. The final indication of inducement is that the practice is discriminatory in that it is not offered to all retailers in the local market on the same terms without business reasons present to justify the difference in treatment.

RED LIGHTS AND GREEN LIGHTS

The federal tied-house law specifically prohibits certain exclusionary practices. These include the resetting of retailer's stock (a supplier cannot touch a competitor's product) and the purchasing or renting of display, shelf, storage, or warehouse space (the supplier cannot pay a slotting allowance). Ownership of less than 100 percent of a retailer is prohibited where such ownership is used to influence purchases of the retailer. A supplier cannot require a retailer to purchase one alcoholic beverage product in order to be able to purchase another alcoholic beverage product at the same time. One consideration on the resetting issue is this: if the supplier is with a wholesaler that represents a wide range of brands, then the wholesaler can do the resetting, within legal limits, without the supplier being in violation. This should also involve significant retailer input to forestall a claim that the retailer's independence has been threatened.

There are a number of practices that are permitted under the tied-house laws, and they are known as "green light practices." These include displays and shelving and other items from which wines are displayed and sold, and which bear conspicuous advertising. In California, displays must be temporary. Also included are inside signs of no intrinsic value to the retailer other than advertising. In California, a sign can have no reference to a specific retailer, and signs in on-premise locations (bars, restaurants, etc.) must not exceed 630 square inches. The federal law permits outside signs, but the California law does not. Wine lists are permitted, but in California they cannot cost more than $25 per unit.

Retail advertising specialties are permitted if they have conspicuous advertising and are primarily valuable for their point-of-sale advertising. They may have some secondary value as well, as a clock would, for instance. In California there is a $50 per brand, per retailer cost limit.

Consumer advertising specialties are permitted if they bear conspicuous advertising and are given to retailers for unconditional distribution to the general public. Examples include ashtrays, wine openers, bags, matches, recipe cards, pamphlets, pencils and pens, postcards, and posters. The California limit is $1 per item and no more than $50 per retailer. Suppliers can provide wine samples to retailers, although in California the wine sample must be one liter or smaller.

Payments for retailer services can be made if they do not put retailer independence at risk, involve interstate commerce, or result in exclusion. California has similar prohibitions but requires no exclusion or interstate ·requirements. California law has exceptions for supplier ownership interests in retailers under special circumstances such as cooking schools, tour boats, and theme parks.

TRADE PRACTICES

Generally, suppliers, wholesalers, and retailers are free to price goods as they like. There are some states where suppliers have to post prices, and then they are required to stick to those price postings. Suppliers can try to influence retail prices, but suppliers cannot control those prices. Suppliers can explain how profitable a particular price would be. They might point out how difficult it would be to reach sales goals unless a specific price was adopted. Suppliers can do things like that but cannot come to an agreement about a retail price or a wholesaler's price. So there are mechanisms to help encourage the pricing at a certain stage, but antitrust law generally prohibits agreeing to resale prices. This includes agreements concerning minimum or maximum price levels.

There are some states in which the supplier cannot provide free goods. An example of a free good is an offer to give one case free if a buyer purchases 10 cases. But this practice is no different than a price reduction that allows the purchase of 11 cases for the former cost of buying 10 cases. Such a price change is not illegal. The State of California Alcoholic Beverage Control Agency agrees that it is all right to have an advertisement that says "Buy two for the price of one" but will prohibit an ad that says "Buy one and get one free," on the basis that the second would be a free good.

Federal law generally governs consumer promotions only if they involve retailers (consumer advertising specialties given to retailers, coupons redeemed by retailers, wine tastings held on retailer premises). It does not govern items and service given directly to consumers without retailer involvement. California law governs "pure" consumer promotions. Suppliers cannot give consumers gifts, premiums, or free goods in connection with the sale or distribution of alcoholic beverages except as permitted by law or regulations. Suppliers may give the general public consumer advertising specialties subject to a $1 per unit cost limitation.

Under the federal law it is unlawful for a supplier to sell or buy wine on consignment, under conditional sale, or with the privilege of return. No exclusion element is required for a violation of the federal consignment provision. An exception is made for bona fide returns for ordinary and commercial reasons arising after the wine is sold. Under California law it is unlawful for a supplier to deliver wine to a retailer where the supplier retains ownership in the wine, or the retailer has the right at any time prior to sale to return the wine to the supplier. Tied-house law prohibits a supplier from requiring a retailer to take and dispose of any quota of wine. Federal law considers tie-in sales to be the same as quota sales and prohibits them. A tie-in sale occurs when a supplier requires a retailer to purchase one product in order to purchase another product, provided interstate commerce and exclusion elements exist.

Federal law prohibits commercial bribery. This conduct involves supplier actions directed to a wholesaler or retailer *employee*, not the wholesaler or retailer entity itself. Commercial bribery includes sales promotion contests sponsored by a supplier that offer prizes to wholesaler or retailer employees, provided interstate and exclusion elements exist. This clause prohibits the payment of money to employees of a wholesaler or retailer without the knowledge of, or consent of, the wholesaler or retailer in return for the employee ordering wine from the supplier. Payments to wholesalers (not their employees) are permitted provided: (1) there is no agreement that the payment will be passed on to employees, and (2) the records of the recipient wholesaler accurately reflect the payment as an asset of the wholesale entity. Such payments to retailers would be considered as inducements under the tied-house law. These regulations do not prevent a supplier from giving money to a wholesaler and the wholesaler deciding to run an incentive contest for employees.

In California, a supplier must keep records of retailer and consumer advertising specialties and samples given to retailers. Federal regulations require a supplier to keep records of product displays, retailer advertising specialties, glassware, tapping accessories, samples coupons, and participation in retailer trade associations.

CONCLUSIONS

Federal and state regulations significantly limit marketing strategies in the wine sector. However, having recognized these limitations, marketing managers find ample room to develop profitable strategies. The laws are a legacy from the repeal of Prohibition, when states were given explicit rights to regulate trade in alcoholic beverages. The laws also reflect the goal of avoiding dominance of the retail sector by producers of alcoholic beverages. There is a difference between federal and state laws, and usually the stricter law applies. The laws define relationships between suppliers, retailers, and wholesalers. Basically they prohibit many activities that are prevalent in other food and nonalcoholic-beverage sectors. These activities include the payment of fees to retailers, ownership of less than 100 percent interest in a retailer, and resetting the retailers' stock. Other activities are limited, including various promotional programs and the provision of wine lists, display materials, and samples.

The summary of certain California and federal laws presented in this chapter is not complete and is not intended as legal advice. For specific applications of these or other laws to any set of circumstances, the marketing manager should seek competent legal consultation. The legal environment is dynamic. Successful marketing strategies need to be dynamic as well, not only to reflect changes in law, regulations, and interpretations, but also to reflect a rapidly changing marketplace.

CHAPTER 8

The Role of a National Importer

Fred Myers

INTRODUCTION

National importers play a key role in the U.S. wine market. They are the conduit through which major import brands are distributed nationally. The objective of this chapter is to raise the understanding of what these firms do and how they implement marketing strategies. The discussion is based on the experience of one major national import company, Frederick Wildman & Sons Ltd., a firm owned by the major brands that it represents.

Chapter 26 discusses the initial determination as to whether an exporting winery should use a national importer or some other organization in entering the U.S. market. This chapter looks in more detail at the national importer option. The first question that should be asked is "Can a business relationship exist between the exporting winery and the importer?" The question may be summarized "Do we fit?" The goal is to determine whether the importer can meet the needs of the winery and whether the winery can meet the needs of the importer. Having competing brands within the importer's portfolio may result in business incompatibility. Differences in expectations about performance will cause problems. Does the winery have goals in the U.S. market that the importer is unlikely to fulfill? Most problems relate to these two aspects of the fit. The same ques-

---

Fred Myers is a 25-year veteran of the wine industry with a background in retail, restaurant, distributor, and importer sales as well as wine education. He is currently vice president and eastern division manager for Frederick Wildman & Sons Ltd., a national importer and marketing company based in New York.

tions form the basis of negotiations between the importer and U.S. distributors. A new importer-producer relationship is delicate, and a series of steps are involved in getting to know each other.

THE FUNCTIONS OF A NATIONAL IMPORTER:
THE WILDMAN EXAMPLE

Wildman uses 30 people to represent and sell its brands on a national basis. This means that one person in the Midwest has eight states to cover and another in the much larger market in Florida may cover two or three large marketing areas. This allows Wildman to present its products in every state.

The importer's job is to keep distributor inventories sufficient for the marketing program and distributor pricing appropriate for the brand and for the market. Keeping adequate inventories in distributor warehouses can be very difficult when demand is surging and supplies lag behind. The importer proposes and assists in sales programs. It cannot direct and can only fund through block grants that give distributors full spending control. It needs to work very closely with the distributor on such programs. The importer develops and carries out marketing programs. Wildman finds that beyond the major brands that it handles, there is insufficient individual volume to support massive marketing campaigns. The entire budget for a brand may be the cost of one full-page ad in the *New York Times Magazine*. The challenge is to take $50,000 and find the very best way to get brand visibility in the market.

One of the most important responsibilities of the importer is to position the brand without a massive investment. The key for the winery is to understand that the relevant brand competition in the United States is different from that in other countries' markets. Therefore it may be necessary to position the brand differently in each of these markets. Riding the wave of an expanding market and moving a high volume of wine do not replace the longer-term need to position the product correctly. Proper positioning should protect profits in declining markets and maximize them in expanding markets. It is a question of identifying the appropriate competitive category, where the brand fits within it, and what activities are necessary to position it there. These issues are discussed thoroughly in Chapters 19, 20, and 21.

Wildman's sole function in order processing is to receive the order from a distributor or wholesaler and communicate it to the relevant export winery. The winery prepares the goods, and the distributor or wholesaler arranges shipment details, although Wildman will arrange them if the buyer prefers. When the goods enter the country, they belong to the customer. While en route, the goods technically belong to Wildman but are covered by insurance. Wildman maintains a warehouse in New Jersey to cover smaller and developing markets and low-volume brands that are purchased once a year. This allows Wildman to allocate prestigious brands more effectively. As brand owner and marketer, Wildman takes a markup on all sales. Other importers cannot buy directly from the suppliers that they represent in the U.S. market.

BRINGING IN A NEW BRAND

A series of hurdles must be overcome in bringing a new brand into the United States, just as in most other markets. The first is to ensure that the supplier and the national importer have agreed on the basic business direction and on clear lines of communication before introducing the brand. The supplier should expect the importer to show proven ability to deliver performance with a brand or a mix of products that fit with the supplier's brand, just as the importer must be assured of a good product fit with its portfolio.

The label for the new brand must meet the strict requirements of the U.S. regulatory agency, the Bureau of Alcohol, Tobacco, and Firearms. It requires information statements to be expressed in a designated form and to appear in a specific label location using a required type size. Other statements on the label are subject to strict scrutiny. The importer has to ensure that all imports are labeled correctly. A few errors can result in an entire shipment being held up in customs. This is not different than in most countries, but it emphasizes the care needed in international marketing because some rules differ throughout the world. The approved label has to be registered or available in every state where wine will be shipped. Most states require that label approvals be filed with the distributor or available from the importer, but many require that each label be registered with the state.

The importer must decide on wholesaler (or distributor) arrangements. It is very important to identify and work with the wholesaler who has the most compatible mix of products and a successful record in dealing with retailers most appropriate for the product. Choices have become more limited because the number of wholesalers has dropped significantly.

The importer has to plan the brand introduction, including kickoff meetings, sales training, and initial programming. This involves ideas for case displays, banners, table tents, and signs. The importer funds many of those projects to the extent that the projects and funding arrangements are legal in each particular state.

The importer also arranges with its suppliers to fund cooperative brand advertising. The importer will match or exceed this budget depending on the brand needs. These funds are spent for marketing materials and for appropriate media advertising (for example, in trade journals or newspaper food pages, or for radio commercials). National advertising may be used, but more often local and regional advertising and promotions are more effective for the promoted brands. Promotion may include involvement with sports events or a program to send wines to critics that decide which wines will succeed or fail.

An important barrier to brand introduction is the lack of sound financing. It is the importer's responsibility to see that its customers, as well as suppliers, are financially sound. This means that the supplier needs to be paid as agreed and must deliver as agreed. The supplier cannot fail to deliver because its growers or bank have not been paid. The importer needs assurance that this will not happen.

PRICING

The hottest market for imports in the United States in recent years was for wines selling at retail for $5 to $10. A price of $20 per case at the exporting winery, ready for shipment, will be subject to considerable markup on its way to the consumer. Table 8–1 provides an example of this.

Notice that the price at retail is more than three times the price at the winery. This spread illustrates the high costs of distribution and also explains the pressures to do away with one level or the other in the distribution chain. Efforts in this regard must recognize that a supplier cannot eliminate the activities that must be performed in distribution. The supplier can change who does the activities and may be able to improve the efficiency with which they are done.

In this example, the price went from $20 per case, the equivalent of about $1.70 per bottle, at the winery in France, Italy, Chile, or Argentina to almost $6 per bottle on the retail shelf and probably $15 in a restaurant. This obviously is not an attractive market for higher-cost producers. As a

Table 8–1 Costs and Margins from the Winery to the Consumer

Cost Classification	U.S. Dollars per Case
Price at winery	20.00
Freight, taxes, and fees	8.70
Importer markup	7.00
Cost to wholesaler	35.70
Wholesaler markup	10.30
Cost to retailer	46.00
Retailer markup	24.00
Cost to consumer	70.00, or about $5.80 per bottle

matter of fact, $20 is already high because some high-volume producers must be selling wine at $14 or $15 per case, FOB the winery. These brands are selling at about $6 per 1.5-liter bottle, and this implies costs that may range between $11 and $15 per case, depending on margins. The market is not delivering a large margin for foreign wineries in this price segment, and that is the reason that California wineries can find so much wine available in bulk. Those bottling it overseas are not making much money and would rather pump it into a tanker than into a bottle.

CONCLUSIONS

The national importer is a link between foreign suppliers and the U.S. distribution network. It acts primarily as a marketing agency in advising on distribution strategies and arranging appropriate distribution links. Through its marketing force, it provides a brand presence in important markets. It develops and executes brand promotions, in collaboration with the supplier, and provides warehousing and shipping service for smaller brands.

The selection of a national importer must be done with careful attention to detail. A foremost goal is to determine that there is a fit between the supplier's brand and the portfolio of brands represented by the importer. Second, the importer should have a proven record of success in dealing with wholesalers that reach targeted retail accounts. Finally, both the supplier and the importer must be financially sound. The choice of a national importer makes sense for those brands seeking national distribution.

Regional importers may be a logical choice for brands seeking more limited distribution.

This chapter has made a clear distinction between marketing and sales. Marketing encompasses the full range of activities that result in a supplier meeting consumer needs at a profit. Sales are that subset of activities revolving around the transaction itself. The national importer helps develop the marketing strategies for the brand in the United States. Wholesalers are the customers of the national importer. They make the sales calls, solicit orders, arrange promotions, make deliveries, and handle the credit and invoicing of their customers. A good national importer will be able to influence these activities in a way that benefits the brand.

CHAPTER 9

Wine on the Internet

Peter Granoff

INTRODUCTION

This chapter considers the principles and mechanics of selling wine on the Internet. The discussion is based on the experience of setting up and operating a single company, Virtual Vineyards (VV), which has become part of a larger Internet-based wine marketing and information company. In the following sections, some references to VV are intended as references to the expanded corporation.

Business on the Internet moves at light speed. Traditional publishing does not. Much of the case history discussed in this chapter reflects the state of VV's business in 1997. Many things have changed since then, with respect to both the Internet and the VV business model, now known as wine.com. While the core value proposition remains unchanged, the technology deployed to support the business allows for much greater breadth and depth. The size of the wine portfolio has expanded dramatically along with the company's network of sources. Warehouse and shipping operations are far more sophisticated, and a European expansion was underway in 2000. As of this writing, wine.com was merging with WineShopper.com, the first major move in a much-predicted consolidation

Peter Granoff was one of the first Americans to become a Master Sommelier. He has served as wine director for San Francisco's highly regarded Square One restaurant and as food and beverage director for the Stanford Court Hotel in San Francisco. In 1994 he cofounded Virtual Vineyards, which is now wine.com, one of the premier on-line retailers of wine.

in the category. In 2000 the company experienced growth of nearly 500 percent over the previous year.

The Internet is a technological phenomenon that is changing the face of marketing. Originally developed as a link among research laboratories to facilitate the sharing of data and results, the Internet became during the 1990s a vast marketplace for goods and services. Some of the services, such as government data series and airline schedules, remain free, but all of the products and many services are sold at a price. This makes it interesting to producers. It is also attractive to consumers. It is open all the time, it is convenient to use, and it offers extensive choice and product information. In other words, it is a logical place to sell wine.

THE SITUATION

As the commercial use of the Internet began to explode after 1990, it became evident that it would be especially useful for selling specialty goods such as premium wines—goods that people tend to want information about at the point of sale. The increasing consolidation among distributors and the growing market power of retailers frustrated small-scale wine producers. These trends made it difficult for small-scale producers to obtain good distributor representation and favorable retailer shelf space.

According to U.S. Census data, in 1950 there were approximately 5,000 alcoholic beverage distributors in the United States. At that time, there were roughly a dozen wineries of consequence in California. By 1996, there were about 250 distributors and over 900 bonded wineries in California, half of them, at least, producing less than 10,000 cases yearly. The two trends ran absolutely counter to each other and left many more producers fighting for representation by far fewer distributors. It is not feasible for large distributors to do an effective job for small wineries, which have difficulties in filling the distributor requirements. Large-scale retailers face similar problems as distributors in dealing with low-volume producers and a proliferation of available brands. In addition, the availability of several wine-rating publications and competitions reduced the need for trained sales personnel. All one needed to do was to paste the right score or medal on the correct shelf space, and the consumer did the rest.

A notable breakdown in the geographic coverage of distributors and retailers occurred at the same time as their concentration and power increased. Many large retailers, served by regional distributors, were selling throughout the country through mail, telephone, fax, and the Internet. As a result, in dealing with regional distributors, wineries were sometimes uncertain how much coverage they would get. It sometimes seemed better, for instance, to sell through a distributor serving a major retailer in Chicago with sales in St. Louis than to sell through a distributor in Missouri.

VV was established to make money by rectifying some of the service losses in the distribution chain for small-scale wineries. It is among that group of retailers called e-retailers—e-tailers for short. VV set out to return that value and expertise to the supply chain by using the Internet. The Internet is well suited for that objective. VV collected more than 80 California wineries in its sales portfolio, in addition to wines from Austria, France, Germany, Italy, Portugal, and the Southern Hemisphere. To gain added sales opportunities, it added a specialty foods portfolio with 50 to 60 suppliers.

The assembly of an adequate inventory to support sales is an important function of marketing. Traditionally, distributors have carried out the assembly function. However, VV took a different track. It developed a consolidated warehouse for its represented wineries. Each winery provides its own inventory, based on an allocation agreement. Sales are generated on the Internet, and the relevant information flows to the warehouse from which the wines are shipped. The VV-designed Web site contains wine offerings, product information, and an order form, and is wired to the inventory database. When the buyer hits the order button, the wine is sold and deleted from inventory, and the credit card is processed. The warehouse packs and ships the wine to the buyer in any amount, from one bottle to multiple cases. Payment is by credit card, as it is with most e-tailers, and credit card security has not been a problem. The margin on each sales dollar must cover Internet, warehouse, shipping, and other sales and administrative costs. The winery makes money if its ex-warehouse price, excluding the e-tailer margin, is properly set.

Wine e-tailers do not fit into the conventional industry categories of broker, distributor, and retailer. In fact, e-tailers do some of the functions of all three of these groups. An e-tailer is similar to a broker in that it does not take title to the wine it sells; it is similar to a distributor in operating a warehouse and handling the shipping and invoicing; and it is similar to a retailer in selling directly to consumers. The e-tailer takes complete responsibility for customer satisfaction. If the customer is not happy with what is shipped, the e-tailer takes responsibility and does not look to the winery for relief.

THE WEB SITE

To be effective as a selling tool, the Web site should be educational, entertaining, and functional. The e-tailer should continually evaluate how well the Web site performs. One characteristic of the Internet is that consumers tend to be very vocal, so a lot of the refinements to the Web site are the direct result of somebody sending an e-mail saying "Hey, you jerks, this isn't working!" VV, for example, decided early on to acknowledge its business objective and its appreciation of comments. It took an editorial position on the Web page to explain the purpose of the site, who the players are, and who has paid for what. Its objective is to make it absolutely clear that the company wants to sell wine and that it is presenting its selection of wines for that purpose.

It is useful to have two layers in the Web site. The top layer can present the editorial message and instructions, while the next layer down provides whatever the wineries have asked to be included. This should not include links to other sites because that encourages customers to "walk out of the store," with no financial incentive for the e-tailer. However, if a winery wants to put recipes, background information on the wine maker, or pictures of the winery's beautiful vineyards, that can be added at no additional cost provided the data are in a form that can be manipulated without added expense.

A properly constructed Web site and supporting systems provide some important sales advantages. Convenience is obviously an important factor; consumers want easy access to a large market with ample information and an easy selling procedure. The Web site can offer personalized services, such as an accessible record of past purchases and their destinations, lists of addresses used, and current shipping information. Internet selling facilitates customer service because it allows for instantaneous communication: the seller can make immediate adjustments or new service offers.

Competition includes other Web sites and, importantly, any good merchant of food and wine that has a good relationship with customers, because buyers consider good traditional retailers with good customer service to be interchangeable with good Internet retailers. Because buyers have many choices, the Internet seller must differentiate its business from more traditional businesses. An obvious strategy is to do something with the technology that a conventional retailer cannot. For example, the VV engineering team wrote proprietary software that provides automatic hourly tracking of all daily shipments by linking to the Federal Express Web site, which offers package tracking. Delivery confirmations received by VV generate e-mails that are automatically sent to the purchasers and include the time of delivery and the name of the person signing for the package. This is an important benefit for persons sending gifts to others, a significant segment of wine buyers.

HOW THE WEB SITE WORKS

An effective Web site will have some navigational headers at the top of the page leading to the home page, the wine portfolio, winery background information, and the order form. There should be an order button accompanying each item; when the button is clicked, the item is "parked" on the form, but the item is not yet ordered and the customer is under no obligation to buy it. Underneath each item, there may be a shelf talker with a chart that describes the wine. In VV's case, this is not a rating but rather information that differentiates the wine from other wine on the site. The descriptive elements are intensity of flavor, degree of sweetness, body, acidity, tannin, oak, and complexity. The elements are displayed as bars, with the length indicating VV's perception of the "value" of each element. This may permit consumers to search the wine selection for desired characteristics.

In addition to the bar-coded description of the wine, the VV Web site was designed to allow the user to click on any one of the major descriptors to obtain a consumer-friendly explanation of what it means. This included an enlarged picture of each wine's label. Other Web sites could have the same or similar features.

MANAGEMENT BENEFITS

Internet sales offer a number of management benefits related to payment and inventory management. The Web site can be programmed to do a preauthorization of a credit card in real time while the customer is still completing the order form. The inventory data bank can be linked to the Web site, to automatically post a sold-out notice beside the item on the

Web site and to notify the inventory staff to pull the wine or restock it. Reorder points can also be established so that when orders deplete inventory to that level, an e-mail message is sent asking the inventory manager to restock. Orders should go directly to the warehouse for pricing and documentation.

Another important tool for managing the business is based on a Web browser interface. For example, while in South Africa speaking at a conference, I found some new wines that looked attractive for the VV portfolio. It was possible to communicate with the Web site directly from South Africa, write up text on the new wines, and put together new samplers. That is the equivalent of a retailer actually going into a stockroom and putting new product on the shelf from 12,000 miles away.

MARKETING AND PUBLIC RELATIONS

One of the big myths about the Internet is that customers come running once a Web site has been established. The myth is far from the truth. The cost of customer acquisition in VV's first year of business was about $175 per person. That was not a sustainable level, and after about two years, the per customer cost was down to about $80. That level requires a high level of sales to be sustainable.

Maintaining customer contact is important for sustaining sales. An e-tailer can use, for example, e-mail to send a bimonthly newsletter that may be sometimes anecdotal, sometimes informational, and almost always persuasive regarding a direct sale. It may announce new wine and food arrivals; they may be seasonal, like sparkling wine and strawberries, or they may be the latest release from a prestige winery. E-mail creates a direct relationship with the customer but requires significant professional support. For example, VV answered 20,000 messages from customers in one year, with all sorts of questions about wine and food that had been raised by newsletters and ads. Perhaps 5 percent of these messages resulted in sales.

It may seem heretical for an e-tailer to use direct mail, but it makes sense to use old media to make people aware of new media. Many potential customers are still in transition from the traditional to the revolutionary. An e-tailer may also buy a presence on other high-volume Web sites. It will have to pay the owners of these Web sites a commission on each sale to a customer that originated from their Web sites. The commission is analogous to a finder's fee. In its shipping packages, VV often included notepads, sales brochures, and a free software CD to facilitate opening an account.

Banner advertisements on the search page are about the only effective way to attract attention in a crowded field. For example, a search on the word "wine" elicited over 13,000 responses several years ago. There is no way that an individual listing is going to get attention. It is possible to buy an advertisement at the head of the list from the searching software firm, for example Infoseek. These banner ads allow an interested searcher to click on the logo and arrive directly at the advertiser's Web site. VV bought the word "wine" at Infoseek, with favorable sales results. Some banner ads produce more consistently than others do; that is why it is important that the Web site design permits an exact tracking of where visitors come from. Otherwise, there is no way to evaluate the results of using banner ads and other strategies to lure visitors to a Web site.

Operating on the Internet requires the patience to educate wineries about the advantages of being listed on the Web site and the skill to educate potential customers about the advantages of buying at the site. This educational role has lessened as e-commerce has expanded, but it still remains important for sales. Although there is a great deal of publicity about the wonderful sales figures some e-tailers are posting, there is still the need to educate individuals and firms about how exactly *they* might benefit. The other issue, of course, is the speed of change in the Internet world. VV's initial software platform was practically outmoded in 2.5 years. It had to be completely redesigned to provide a more robust database and better cross-merchandising possibilities. What the consumer really wants is to put a glass right up to the modem port and fill it with wine. Until this is possible, attention is focusing on maintaining and improving the speed of responsiveness to consumers in everything from delivery to answering e-mail queries.

REGULATORY OBSTACLES

The wine trade is subject to regulation in all countries, but the regulations are perhaps most complex in the United States, where each state has the right to regulate alcohol production and marketing in any way it sees fit. Sellers must work hard to ensure that their shipments comply with applicable state laws. In some cases direct shipment to consumers is prohibited. Like wine retailers, wine e-tailers must be fully aware of the various regulations affecting operation of the business.

The other side of this issue is that Internet selling is helping gradually erode the geographically based regulations that hamper trade. This erosion arises partly from the difficulty with effectively maintaining regulations in the face of profitable commercial opportunities, and partly from consumers demanding access to modern and convenient ways to shop.

There are entire bureaucracies responsible for the regulation of industries (such as wine) that operate on the assumption that the physical locus of a commercial transaction is fixed. A fixed place defines the geographical area of jurisdiction and permits more readily enforceable tax rules as well as regulations of business hours and other conditions of operation. The Internet has thrown into question the concept of the place of transaction. An example from Japan illustrates the point. Japan has different regulations controlling the issuance of credit cards than the United States does. In the United States, there are merchant bankers that can process anybody's credit card if they want. There are intermediaries that actually develop and operate much of the technology and deal with the merchant banks. It is illegal in Japan for a bank to process a credit card on the Internet. One of the big trading companies in Japan actually has a very successful Web site marketing products to Japanese consumers, but it is illegal for the company to actually process the credit card transaction in Japan. Its solution is to hand the transaction across to a computer in California, where it is processed through a merchant bank and returned to Japan. The process takes about 15 seconds, and no law is broken. How long can such restrictive regulations exist in Japan when the Internet offers an efficient and legal alternative? The point is not to encourage businesses to become outlaws but rather to recognize the need to change the regulatory structure to meet the needs of a technologically advanced society.

CONCLUSIONS

In the Internet, wine producers that have been squeezed out by the large-scale traditional distribution structure have a new place to market their product. Internet sellers defy traditional classification. They perform some of the functions of brokers, distributors, and retailers. Typically an Internet seller will not take title to the wine it sells; it will operate a warehouse and handle shipping and invoicing. It will sell directly to consumers and take responsibility for the products that it sells.

An Internet seller differentiates itself on the basis of technology. The e-tailer must have a properly constructed and easy-to-use Web site that provides convenience, personalized services, and good value. It also should provide integrated inventory control, shipping instructions, and necessary documentation. Competition includes other Web sites and good merchants maintaining strong customer relationships.

Operating on the Internet requires an educational effort directed toward both suppliers and consumers. It also requires the ability to keep up with rapid technological change. Finally, it requires the ability to comply with laws and regulations that exist in the varied jurisdictions covered by the Internet. In the final analysis, success will depend on how well consumer needs are met at a price that will satisfy the consumer and yield a profit for the seller.

PART III

Developing Strategies

CHAPTER 10

Pouring Wine
through New Funnels

Bruce H. Rector

INTRODUCTION

Why is it so hard to sell wine? Because wineries are all pouring themselves through the same funnels. Funnels are brand concepts or unique selling propositions—marketing ideas. This chapter briefly examines the approach to developing new ideas or funnels. I've developed this approach in collaboration with colleagues and clients—the "we" of this chapter. Its purpose is to make marketers understand the very basic ideas that underlie innovation. The focus is on the process of thinking rather than on the idea itself.

THE FAMILIAR FUNNELS

Before attempting to innovate, one should understand the funnels commonly used right now. The first funnel is *appellation*, the concept that ties the wine uniquely to its origin. It works well for some wines and appears irrelevant for others. The idea is that if you want this flavor, it comes only from this place or vineyard or soil. The second funnel is *family*. This ties the wine's character and reputation to the character and reputation of a

---

Bruce H. Rector received his enology degree from UC Davis, gained production experience with several wineries, and became the wine maker and a partner with the Benziger family, helping them grow the Glen Ellen brand to four million cases. Following the sale of the Glen Ellen brand, Bruce has focused on the spiritual side of wine making, developing the notion of "Harmonic Wine Growing" and creating his own brand, Abundance Vineyards.

family. Its objective is to build confidence among consumers that the brand is about families and their truthfulness and integrity. Examples of wineries using this funnel in the United States are Gallo, Mondavi, Sebastiani, Kendall-Jackson, and Wente. The next funnel is *price*. It includes (1) selling on the basis of high price and product scarcity, (2) selling at a low price to grab market share, and (3) selling on the basis of value. In the third case, the wine has to be a good value for the price, or the consumer will not buy again. The fourth funnel is *visual appeal*. This is based on developing an attractive label, bottle, and box and supporting them with appealing visual images. Most wineries use a combination of these funnels in their marketing strategies.

LESS TRADITIONAL FUNNELS

Escape from tradition has led to some less frequently recognized or used funnels. One is the funnel of *process*. This funnel emphasizes some unique process involved in making the wine that makes the wine special and perhaps appealing to a particular market segment. The organic wine business uses this funnel. Another less traditional funnel is *function:* the wine or its package provides a function that is valued by some buyers. One example is sacramental wine. Another is bag in the box; some customers like the box because it fits easily in the refrigerator, it does not spoil, and it is easy to use. Third, there is the funnel of *humor* or *antithesis*. A lot of Bonny Doon and Barefoot Cellars wines are marketed with humor, tending to parody the traditional funnels.

Most wineries overlook the most important funnel of wine marketing. When people sit down for dinner and open a bottle of wine, what often happens to them? Their conversation becomes more intimate and they begin exploring topics they might not otherwise consider. So wine has an intrinsic psychoactive power, but wineries do not really think and talk about it. One of my efforts is to create brands that use the relationship funnel, brands that convey the idea of the stimulating experiences that happen when friends are drinking wine and immersed in good conversation.

Another funnel or brand concept is *charity*. We have a brand called Abundance, which features a chardonnay and a sangiovese that sell for $9.99, a good price point for them. We could not decide on a charity that would not alienate anyone. So we decided to let the customers decide where this money should go. Customers soak off the back labels, then send them to any charities that do not advocate violence. Charities consolidate the labels, send them to us, and in return receive $2 for each soak-off label. Although this is a small brand, we have already given away over $5,000. Schools and theatrical groups and Kiwanis Clubs love this idea. People are going into stores to ask for the wine in order to help their charity. That helps the distributor and helps us.

Not all nontraditional funnels work. We had a concept called Cuisine Cellars in which we sought to build on the relationship between wine and food and the observation that people have little time to prepare food. On the back of the label we put a tested recipe guaranteed to go with the wine in that bottle, figuring that people would welcome the easy, fast, and delicious meal idea. Unfortunately, we overfunctionalized the label. We lost the romance of the wine.

A really simple way to develop funnel ideas is to visit a supermarket where wine is important and read supermarket trade journals. Look for trends in consumer in-store behavior; identify which aisles are shrinking and which aisles are expanding. Then try to figure out how to take advantage of those trends and into the expanding aisle. Of course, the concept has to fit with your overall marketing identity to be credible.

COMMUNICATING THE IDEA

Wineries need to innovate with an eye to what consumers will accept and to communicate effectively about the concepts that are adopted. The critical consideration in adopting a concept is whether the concept is ahead, behind, or on the power curve of consumer thinking. Wineries should design so that energy comes toward them. If an idea is too far ahead of the power curve, many people will complain that they do not "get it." Worse, they will simply ignore the idea. An idea behind the power curve will not work any better. Wineries should aim to be at the leading edge of the power curve, where the energy will come toward them. This takes trial and error, unfortunately, and wineries must decide how many people they want to appeal to. When developing ideas, wine marketers should seek out assistance from colleagues, friends, or acquaintances who can stimulate thought, compare ideas, and be honest critics.

Marketing is about storytelling. Wineries need to effectively tell the story about the new idea or concept in a way that people will relate to. Think of the long tradition of storytelling that has survived centuries of print and decades of electronic media. Wineries can follow traditional storytelling models to tell the story of how they came to be and what they are doing.

CONCLUSIONS

The funnel theory is just another way to look at marketing strategies. Developing new funnels is a way for wineries to distinguish themselves in order to survive in the marketplace. Our experience with new funnels linking wine to charity, food, and intimate friendships has shown that marketing innovation involves trial and error but can ultimately bring great success.

CHAPTER 11

Product Differentiation

Thomas H. Shelton

INTRODUCTION

The successful wine marketer will sell the last bottle of wine at the highest possible price on the day before the succeeding vintage is released. The challenge is to figure out how to become a successful wine marketer in the first place. The purpose of this chapter is to explore this question with a particular emphasis on how to differentiate a product and create market niches.

PRODUCT-DRIVEN AND MARKET-DRIVEN WINERIES

It is very important that wineries understand who they are and what their goals are. I think that there are two basic identities: a product-driven winery and a market-driven winery. Table 11–1 lists the characteristics of these two types.

The product-driven winery focuses on making the very best product it can and differentiating its brand on the basis of quality. Product quality considerations will dictate how the winery invests in vineyards and selects tanks, pumps, and barrels. These decisions tend to make the winery a

---

Thomas H. Shelton has more than 25 years of experience in wine sales and marketing, including positions with Vintage Wine Merchants and as vice president for sales at Franciscan and Guenoc wineries. He is president and CEO of Joseph Phelps Vineyards and chairman of Free the Grapes, an industry group working to reform the current three-tier distribution system.

Table 11–1 Characteristics of Product- and Market-Driven Wineries

Product-Driven Wineries	Market-Driven Wineries
High-cost producers	Low-cost producers
Price-resistant products	Price-sensitive products
High perceived quality	Value orientation
Recognized origins	General origins
Production limited by quality	Production limited by demand
Quality/brand relationship	Price/brand relationship
Low sales and marketing costs	High sales and marketing costs

high-cost producer. The responsibility of marketers is to create or build a profitable market for that wine.

Market-driven wineries are another story. They study market trends and try to create products that exploit those trends. Outstanding examples exist in California of wineries that capitalized on the growth in white zinfandel, chardonnay, and merlot. Their sense of the market and how consumer preferences will move in the future guide their investment and production decisions. Market-driven wineries by nature have to be more sensitive to costs because their product is generally going to be much more price sensitive. For instance, when there was a shortage of premium grapes in California, a lot of wineries bought wine from France, Italy, and Chile and marketed it under brands that have traditionally been for California wines.

Alternative sourcing happens first among market-driven producers when grape prices increase because producers are looking for wines that allow them to remain competitive at their target price points. For example, if a winery believes that $7.99 is an important consumer price point and higher grape prices will make that point unprofitable, the winery will look for less expensive fruit or wine from other areas.

The consolidation of distributors has made marketing more difficult than it was when a winery had considerable choice among various outlets for wine. Today, most wineries need more clout to get attention and service. This led my winery, Joseph Phelps, a product-driven organization, to combine its marketing with a larger concern called Co-Brand Corporation, a collection of cooperating wineries seeking greater influence at the distributor level. As effective and strong as Co-Brand is, with 100 salespeople on the street nationally, we are still between 10th and 15th in terms of importance to distributors. We end up fighting for the scraps that are left over on retailer and restaurant lists after these distributors have placed what they consider to be their mandatory brands.

SUSTAINED PRODUCT DIFFERENTIATION

Product-driven wineries need to build a big brand image if they are to compete in the new distribution structure. The purpose is to pull wine through the distribution system using product differentiation and niche marketing. Brands can be differentiated through packaging, blending, varietal content, number of awards, "star" wine makers, processes used, quality of equipment used, and origin of the grapes. All of these except the grapes' origin are nonsustainable. Competitors can eventually match what the winery is doing. These differentiation strategies need continual attention and development.

There is a hierarchy of differentiation that a winery can move through as its product class becomes smaller. The hierarchy begins with brand differentiation and ends with a proprietary name with estate designation, as illustrated in Figure 11–1.

More California producers are beginning to see the wisdom of the French concept of appellations because they provide sustained differentiation. The difference is that the French appellations include controls on varieties, yields, and other matters that are, so far, unpopular with California producers. However, individual wineries, or groups of wineries, may adopt similar controls in dealing with grape producers in order to gain differentiation advantages. Appellations like Napa Valley and Sonoma County, and various viticultural districts and individual vineyards, are examples of this trend toward sustainable differentiation. If the viticultural district provides valued quality attributes, then it confers an advantage over competing wines from other areas. Over time, however, competition among viticultural districts is likely to result in the development of cultural or other rules that will further distinguish one district from another.

Vineyard designation is a more specific differentiation strategy that is sustainable if the winery owns the vineyard. Phelps has been associated with two well-known Napa Valley vineyards, Eisle in Calistoga and Bacchus in Oakville. We leased the vineyard in Calistoga for 20 years and were very influential in building its reputation. When it came up for sale in

BRAND
(One for all, or many for one)

PACKAGE
(Label, bottle, size, box)

VARIETAL
(One, many, or unique)

STAR WINE MAKER
(Wine dinners and public relations)

IMAGE APPELLATION
(District or vineyard)

ESTATE DESIGNATION
(Exclusively a winery's, but can be copied)

PROPRIETARY NAME
(Hard to beat, but others exist)

PROPRIETARY NAME WITH ESTATE DESIGNATION
(About as exclusive as possible)

Figure 11–1 A Hierarchy of Differentiation for a Winery as Product Class Becomes Smaller

1991, its high price reflected much of our marketing efforts over the years, and we did not buy it. Thus, we lost a source of competitive advantage. We did not make the same mistake in Oakville. When that vineyard came up for sale after we had leased it for years, we bought it and continued to gain the advantage of its high-quality fruit.

This strategy allows the winery to produce estate-bottled wines. This is a very valuable concept that is underutilized because it restricts sourcing and blending opportunities. Once the estate concept gets going, however, it does provide an exclusive way of differentiation that is entirely under winery control. The production of "reserve" wines is another means for differentiation, but it is not sustainable. Any winery can have a reserve wine, and there are no legal rules that define the term in a meaningful way. There is no consensus within the industry as to what "reserve" means.

However, it can be an effective differentiation strategy for a strong brand with a high-quality image, such as Mondavi or Beaulieu.

USING A PROPRIETARY LABEL

A proprietary label provides another way to differentiate a wine. Phelps was among the first in California to do this with the trademarked name "Insignia." In our case, we used the mark to designate the best red wine we could make regardless of variety, origin, or blend. Insignia currently represents something less than 7 percent of our total output. In the first year, 1974, it was cabernet sauvignon from Stags Leap. The next year it was merlot from Stanton Vineyards. This was followed by a cabernet from Eisle vineyard, and we sacrificed the vineyard name to strengthen the proprietary concept. In subsequent years, the best wines were blends rather than individual varieties, and Insignia ended up covering the best Bordeaux-type wines that incorporated cabernet sauvignon, merlot, and cabernet franc. Other varieties, such as petit verdot, may be added as their quality warrants. The wine is labeled as Insignia Red Table Wine, with the varietal content listed in small print below the name.

Although Phelps owns the name, it does not own the concept. A fairly large number of premium wineries belong to the Meritage group and produce Bordeaux blends under that name. They are trying to duplicate the success that we have had with Insignia. Like Insignia, Opus One is a proprietary concept and has had great success. The fact that the value of the proprietary name can be eroded through competition means that wineries need to continually develop new ways to distinguish their wines.

Phelps discovered that fruit from Stags Leap and from Rutherford Bench consistently produced the best Insignia wines. We began to narrow our sources and eventually purchased equal-sized vineyards in each of these two areas. This permitted the wine to be both proprietary and estate. I believe that there is no one else in the world that can duplicate Insignia without coming into our vineyards and harvesting the grapes in the middle of the night. In terms of differentiation, Phelps has achieved brand recognition, a trademarked proprietary name for a quality wine, and a locked-in source that no one else can tap. In my view, this is the best way to protect a brand in the marketplace and completely differentiate it from the competition. The differentiation process has been profitable. Insignia used to sell for $25 a bottle, but today it is selling for $70, which is $20 to $30 below our target price. We are trying to establish a benchmark price for Insignia based on the prices for Bordeaux wines and those of California marks such as Dominus, Rubicon, and Opus One.

OTHER DIFFERENTIATION STRATEGIES

We also hope to differentiate Phelps' wines by pricing to vintage, a concept we began to explore in 1997. Price will be based on our assessment of the quality of that vintage, with the possibility of reducing prices from one vintage to the next when we think quality is lower. The success of the strategy depends on consumers' knowledge of variations in quality and their willingness to accept price variability over price stability. There is no question that California premium wines vary in quality from vintage to vintage, and this is more noticeable among expensive wines drawn from very narrow producing districts and vineyards.

One of the more successful differentiation strategies we have used is the establishment of a brand with an attractive French-style image. The name is Vin de Mistral: the wine of the wind that blows through the rich Rhone Valley. We believe that the name conjures up a romantic image associated with that valley, although others argue that the mistral is anything but romantic. Our objective is to market a style of wine that is readily recognized and appreciated by a particular group of consumers who are tired of varietal monotony. Within that brand we produce a wine called Le Mistral, which is a California interpretation of the Rhone reds, that is, a blend of grenache, mourvèdre, syrah, and other grape varieties. We also produce grenache rose, syrah, and viognier under the same brand. The principle of this strategy is to combine a real product difference with an image difference. We are not just creating a new image for Phelps; we are creating a different style of wine and positioning it with a specific product image.

CONCLUSIONS

Successful marketing involves a mix of strategies that work together to produce sales at a long-term profit. This chapter has focused on one strategy among them, product differentiation, and suggested how this strategy might differ for market- and product-driven wineries. It emphasized advantages that can be gained through the use of a proprietary name for the wine product and described a strategy involving origin, product, name, and brand differentiation that has been successful for a midsized California winery. The lessons from this experience are that marketing needs to be dynamic because few differentiation strategies are sustainable over the medium or long term; wineries should seek more sustainable strategies; appellations of origin concepts are increasingly viable as competition increases; and strategies that combine differences in product, brand, and image, rather than focusing on just one difference, can be very effective. These are the strategies that help establish a sustainable competitive advantage.

CHAPTER 12

Building a Premium Wine Brand

Agustin Francisco Huneeus

INTRODUCTION

Brand marketing in the wine industry is a relatively new phenomenon that is related to the rise of New World producers. In the Old World, wine was marketed and "branded" by its place. Today, brands often have little or nothing to do with place, and everything to do with factors such as image, packaging, and positioning.

Building a brand simply means making it stand out among its competitors so that consumers recognize and buy it. This requires marketers to differentiate a brand by using a variety of strategies. For wine, building a distinct brand can include strategies that are not always apparent on the label. This may include production and blending decisions, pricing and distribution strategies, and other sales and promotion tactics. Differentiating a brand also includes indicators that do appear on the label. For example, geographic origin of the wine (which may also imply some specific production techniques), the wine's variety or combination of varieties, and its designation as an estate or proprietor wine, all serve to differentiate a brand. Another more recent phenomenon of brand building is the creation of ultra high-end proprietary blends with proprietary names such as Opus One, Insignia, Magnificat. All of these brand-building strategies must be

Agustin Francisco Huneeus is a graduate of UC Berkeley and received an MBA from Northwestern University's Kellogg Graduate School of Management. After working with the venture capital group of Hambrecht & Quist evaluating branded consumer companies, he joined Francisan Estates as senior vice president of sales and marketing in 1996, and was promoted to president in 1999.

an integral part of an overall marketing plan that includes promotion, advertising, sales, pricing, and distribution.

THE SIGNIFICANCE OF WINE CATEGORIES

It is a sad truth that regardless of how incompetent a winery might be as a marketer or a brand builder, if a wine is marketed in the right category, it will succeed. For example, a winery making a Napa Valley merlot in the late 1990s could hardly fail, even with inept marketing and an unattractive label. Its wine would be on allocation (rationed to buyers), not because of the brand but because the brand is in a "hot" category. Of course, a skilled winery that capitalizes on a growing category can do very well, as illustrated by Kendall-Jackson in the chardonnay category.

Being in the wrong category can thwart a successful marketer just as easily. For example, Peter Sichel, the very successful developer of the powerful Blue Nun label in the U.S. market, was unable to maintain a market for his riesling wines because the entire category fell out of favor with consumers. The market for rieslings became very limited as other categories like chardonnay and white zinfandel grew in popularity.

It is possible to have a good brand in a dying category as long as volumes are sufficient to be profitable. Constellation Brands, the parent company of Franciscan Estates, provides an example of this, with brands such as Almaden in the bag-in-the-box and jug wine categories. They are fantastic brands within their given categories, but they have still declined over time because, quite simply, their categories declined. They are not the wines that most people want today.

I work for the Franciscan brand. Franciscan's strategy is based on balancing category, brand, and origin in a way to maximize returns. This has led us to structure the company a little bit differently. Whereas many California wineries, especially of Franciscan's size, tend to buy grapes from several growing areas and use brand and variety as the principal differentiation strategies, we stress vineyard origin and brand.

Franciscan believes that the single most important characteristic of any given wine is the soil from which it comes. Therefore, everything that we make is estate based, even though we could sell far higher volumes of merlot, for example, if we sourced grapes from other places. Franciscan wines come from the Oakville estate only. This has allowed us to develop the Franciscan brand from a relatively weak position 10 years ago into a brand today that is totally allocated, has almost 70 percent on-premise distribution, and is a distributor favorite because it moves quickly at a full markup and sells itself.

COMPETITION AND EVOLVING CHANNELS

Competition is very intense in every category and at every level in the wine market. This means that the brand needs to capture attention in its price category and among distributors, retailers, and consumers. In 1976 there were 240 wineries in California; in 1986, there were 600; and by 1999, there were over 800. On the other hand, the numbers of California distributors went down from 28 to 15 to 5. This means that, on average, there are 160 California wineries per distributor. The range of representation is large, with the biggest distributor accounting for approximately 250 wineries. Consequently, each winery has to fight for attention and devise new strategies to exercise marketing clout so that the distributor's sales force is motivated to present the brand to target retailers. The biggest distributor in California, Southern Wine and Spirits, carries all the big players, such as Beringer, Meridian, Mondavi, and Clos du Bois. So when a smaller wine marketer like Franciscan Estates seeks representation in expanding a smaller, estate-based brand like Estancia, the response may be no more than a stifled yawn. That is why a winery needs a sound marketing strategy and the elements of a very strong brand.

Retail channels are also consolidating. Twenty years ago, retail distribution was mostly in the hands of independent retailers with one or two stores in their market. Wine ratings were less prevalent, and the retailers seemed more likely to use their own judgment in selecting the wines that they offered. Since that time, supermarket chains, discount chains, and large marketing clubs have gained significantly more marketing power.

Restaurants and hotels have also moved away from individual location buying, even for chains, and are increasingly using centralized purchasing to gain bargaining power and the consequent economies of scale. Purchasing decisions are often made by buyers with inadequate experience in wine, little interest in the "vineyard story," and a sole focus on price, product turnover rates, and response to advertising efforts.

Clubs have become very strong in some markets. Their competitive strategies often are based on a narrow range of very fast moving wines. They may have two or three chardonnays in the $7 to $8 price range, for example, rather than 10 or 15, as would be found in a supermarket or other wine store. A related type of discount store is the "category killer," which focuses on a single category and carries every line item in that category at a discounted price. These are large stores, working on low overheads and markups that are generally below those of competing outlets. This is a popular retail concept used in stores offering office supplies, building supplies, and consumer items.

The consequence of this change in wholesale and retail structure is that wineries have fewer outlets to sell to and these outlets prefer to deal with strong brands that have more market clout, more money to invest in brand development, and extensive advertising and promotion support. The characteristics of a strong brand vary according to the price category of the wine, but strong brands have the common ability to pull, rather than push, product through the marketing chain. Examples of California brands with pull include Gallo premium wines, Robert Mondavi, Kendall-Jackson, and Silver Oak.

MEASURING BRAND STRENGTH

A successful brand strategy creates a situation in which buyers pull the brand through the market rather than relying solely on producer push. There is a story about a restaurant owner who did not want to do business with the distributor of the Cakebread Cellars brand. No matter what the owner did, he could not bury the Cakebread brand in his wine list. People kept asking for it, and he had to carry it. This is an example of brand pull, one measure of a strong brand.

At Franciscan Estates, we use InfoScan data to measure brand strength because it allows us to calculate market turns for various competitors in supermarkets. Data for 1996, for example, showed that Kendall-Jackson outsold the chardonnay competition, selling 10 times the volume of the next brand and 30 times the volume of Franciscan. Analysts can also compare wineries along two axes, using the average producer (winery) price and the volume of sales (in cases). The strongest brands are able to sell more wine than competitors in the same price range and at a higher price than competitors with similar sales volumes. This sort of comparison shows that a number of wineries, such as Gallo, Mondavi, Kendall-Jackson, and Jordan, have stronger brands than their competitors. If a winery can sell 100,000 cases of wine at $100, it is in the "pack" of brands; the strong ones will be able to sell at $150 or $200. And then there are the really powerful brands with very low average prices but very high volume. Gallo, as the strongest brand in this size category, is able to sell at higher average prices than its competitors.

The Gallo family of brands also has strength through the winery's sheer size, representation in every category, tremendous name recognition supported through extensive advertising and promotion, a large and effective sales force, and strong distribution. They market wines produced in most California regions as well as other countries.

Robert Mondavi has brands in all the premium segments, from the popular-priced Woodbridge to the ultrapremium Opus One. The brands have excellent consumer recognition and tend to sell themselves. With this brand strength, Mondavi successfully markets wines from wine-growing regions in California, Chile, Italy, and France. The company has tremendous power in the distribution chain because it is a brand leader and solid moneymaker for distributors.

Kendall-Jackson has developed a very strong brand and formidable distribution by concentrating on a single product, Vintner's Reserve Chardonnay, sold in the high-premium price class ($7 and over). By 1998, the Kendall-Jackson brand had become dominant in this price class. As part of its focused strategy, the winery positioned its affiliated brands in this market as well. The winery has been in a fast-growing category over a long enough time to develop an enviable reputation for consistent quality and value, and a very high turnover for the product. This, in turn, attracts distributors and allows them to take a smaller margin on Kendall-Jackson than for slower-moving brands. It gives the winery some leverage in raising its prices without seeing a corresponding rise at the retail level.

Silver Oak developed brand strength by concentrating on a single variety, cabernet sauvignon, in the luxury premium class (wines selling at over $25 per bottle). The total output is small relative to that of many others, and the wine always appears to be in short supply. The aura of an expensive wine in short supply tends to stimulate demand and maintain prices at a profitable level. The wine virtually sells itself.

BUILDING BRANDS

The building of brands is more difficult than it was a decade ago because there are more wine brands available and fewer distributors to sell them. Consumers have more wine brand choices than ever before. Consequently, the use of vineyard designations, proprietary names, and "favorable image" brand names has become much more important in the marketing of premium wines. Franciscan Estates combines the basic differentiation strategies of brand and appellation or vineyard designation to position the product as a prestige wine in limited supply. The objective is to develop brand "pull" so that we do not have to push the wine through the market.

Franciscan has vineyards in Oakville, noted as the origin of critically acclaimed and expensive wines. The premier brand is Franciscan Oakville Estate. From this estate, we make cabernet sauvignon, merlot, zinfandel, a proprietary red blend called Magnificat, and a very high-end wild-yeast chardonnay. Thus, we have chosen to offer a range of varieties in the high-premium price class rather than concentrate on a single variety, as has Kendall-Jackson or Silver Oak.

We are taking advantage of what our vineyards will produce and the power of a vineyard designation to sell it. We offer what no other competitor can, the attraction of the Oakville vineyard designation and the trust of the Franciscan name. The strategy works for us, and a similar strategy works for Robert Mondavi. By nurturing demand ahead of supply, we built a strong brand that is totally allocated and profitable for distributors.

In 1986, Franciscan Estates also developed a lower-priced brand, Estancia, to compete with some of the higher-volume producers. Estancia began with estate vineyards in Alexander Valley and Monterey County. White wines and pinot noir come from Monterey, and the red wines come from Alexander Valley. Estancia Cabernet is sold in the superpremium price class ($10 to $16) and is an Alexander Valley cabernet sauvignon, coming from the same area as the more expensive Jordan and Murphy-Goode wines. Estancia Chardonnay competes with very large brands but creates value because it is from the Pinnacles ranch, a recognized vineyard that you can walk across, gaze upon, and feel.

Franciscan Estates also uses proprietary names to differentiate some of its wines and reduce the company's exposure to market losses as varietally designated wines become more like commodities. The most frequently used name for California-made proprietary blends is Meritage. The Meritage Society, partly founded by Franciscan, authorizes the use of the name "Meritage" to designate high-quality blended wines that cannot legally be defined as a single variety. Franciscan Estates makes several

Meritage red wines, which are blends in different proportions that may contain cabernet sauvignon and cabernet franc. The names for these blends are Magnificat for blends from Franciscan Oakville Estate, Red Meritage for the Estancia estate in Alexander Valley, and Mount Veeder Reserve for the Mount Veeder estate. These names provide a distinction for these wines and estates in addition to that gained from vineyard and varietal designation.

PRICING, PRODUCTION, AND DISTRIBUTION

Building a premium brand requires careful decisions about product pricing, production levels, and distribution strategies. Price is closely related to image. Many people evaluate wines based on price: the higher the price, the better the wine must be. Conversely, lower prices are associated with lower quality. A prestigious brand is incompatible with low prices and mass distribution. It must deliver value over time to sustain its price and limited production.

To build the brands within the Franciscan Estates portfolio, we had to make wines at the "right" volume and sell them at the "right" price in the "right" place. Our conclusion is that if a winery wants to build a great chardonnay image, it cannot sell just in Safeway but must be in recognized wine stores and celebrated restaurants. Wine is among the few consumer products for which the distribution profile is an integral and crucial element of the marketing mix. With most products, distribution does not affect marketing. One does not think any less of Coca-Cola because of where it was purchased.

The distribution strategy also has to account for channel conflict. For example, a large brand may want to be in a major supermarket chain. This decision, however, may cut the brand off from independent retailers that object to the chain's feature advertising at low prices. The supermarket chain advertises the brand at $8.99 and a key restaurant, the "image" account for the brand, decides to remove the wine from the wine list because the restaurant has been selling it for $6 per glass and customers have begun to object. The chains object when the brand appears in clubs and discount stores that work on a 7 percent to 19 percent margin as compared to 30 percent for the chain. This sort of channel conflict leads to the development of different brands, each targeted for a special sort of outlet.

CONCLUSIONS

Brands have become very powerful over the past decade in the wine business. This has allowed brand owners to expand their sources for wine and capitalize on their brand recognition to sell the wine. Franciscan Estates sells wines from Napa Valley, Sonoma Valley, Chile, and the Central Coast; Mondavi sells wines from Napa Valley, the Central Valley, and France; Gallo has done it with wines from Sonoma County, San Joaquin Valley, and Italy; Fetzer sells wines from the North Coast and Chile. These brands constitute a wine marketing force.

In my view, there are three ways to succeed in building a brand. The first is to have distribution clout, as do Gallo, Mondavi, and Kendall-Jackson. These brand owners can virtually dictate to their distributors where they want to be sold. The second is to have a brand that sells itself, such as Silver Oak. The winery produces the best wine it can, puts a premium price on it, and the customers just keep coming. The third way is for the winery to go out and sell the wine itself, creating the illusion of scarcity to further increase demand. This is the course followed by most wineries that lack the size, clout, or tradition of the big guys. It means selling the wine and requires marketers to make all the right decisions on pricing, packaging, promotion, distribution, and more, so that the product is pulled through the distribution chain.

A story about a friend of mine illustrates why brands have become so powerful in the premium wine sector. My friend knows little about wine and wanted to buy a bottle for his father-in-law. The clerk asked what kind of wine he was looking for, red or white. My friend said that he was not sure, but he went ahead and picked white. The clerk said, "We've got sauvignon blanc, chardonnay, riesling, Alsatian wines," and so on. My friend opted for chardonnay. "OK," said the clerk. "We've got 30 from California, 20 from France, 15 from Italy, 4 from Chile, 5 from Australia, 4 from New Zealand, and 6 from South Africa. What are you looking for?" At this point, I would have been ready to run out the door and buy some flowers. The dizzying array of choices intimidates customers. Now, with the growth of brands, when people scan the wine shelves and spot a familiar logo like the Kendall-Jackson gold leaf, they get a tremendous feeling of security. Regardless of the source, they are confident that the wine will be OK and the label will be recognized when the wine is served. This is a definition of a strong brand.

CHAPTER 13

Pricing and Programming

James Cahill

INTRODUCTION

Pricing and programming are two critical elements of marketing strategy. Pricing is easy to think about but difficult to implement. Programming is the scheduling of various price-related actions such as discounts, incentives, advertising campaigns, and new vintage or new product releases. Pricing decisions are influenced by profit objectives, brand competition, conventional pricing practices common to individual markets, various legal requirements, the relationship between everyday and feature prices, and the type of the outlets targeted for sales. Programming requires working with distributors and other buyers and involves a calendar that schedules monetary allowances for special purchases, quantity orders, and inventory depletion. It may include monies for ads, floor displays, new distribution, wine-by-the-glass programs and other promotional activities, as part of a sales incentive program. The examples cited here are mostly from the U.S. market, but the principles developed apply to pricing and programming strategies globally.

James Cahill received his MS in enology from UC Davis and has worked as vice president of marketing at Round Hill and as divisional vice president of Austin, Nichols and Co. He is vice president of sales for Supreme Corq, Inc., a supplier of synthetic corks to the wine industry.

ESTABLISHING A RETAIL PRICE

Wineries can target retail pricing points that they believe will be most advantageous for selling their products. However, they can rarely dictate those prices. Therefore, they need to establish the conditions that will make the targeted pricing points attractive for those selling the wine. Expected profit per case is one consideration in identifying a feasible retail pricing point for the producer. The profit calculation should reflect costs, output, and realistic expectations about pricing, given the market situation and the nature of the product. Once a profit objective has been established, the firm can determine how well various pricing strategies might achieve the objective. This point cannot be overemphasized, because it is relatively easy to assess how pricing might affect gross revenues but more difficult to evaluate the impact on profit. Thus, a firm should focus on the effects that incremental shifts in retail price points are likely to have on profits per case rather than just on gross revenue. From this information the firm can determine how many cases must be moved, at various feasible price levels, to achieve or exceed profit objectives. The profit definition used here is gross profit, before accounting for sales, marketing, and various overhead costs. A more rigorous analysis would examine impacts on net profits.

Prices and costs are variable because wine is an agricultural product and its availability varies over time and from market to market. Rarely is there just the right supply to serve market demand at a profit; therefore, prices and profits must be adjusted. The following example illustrates how the gross revenue objective leads to different adjustment decisions than the profit objective. Consider a product with production costs of $35.40 per case, a winery price (FOB price) of $56.00, and a retail price of $9.99 per bottle, based on customary wholesale and retail markups. Monthly shipments are 1,000 cases. In the initial period, gross revenue is $56,000 and profit is $20,600. If production is to be increased, and if the firm has a strategy to maintain gross revenue, then a retail price of $6.99 will be set to move 1,500 cases per month, based on assumed market conditions. This will maintain gross revenue but leave a profit margin of only 5 percent as compared to 37 percent earned previously. This strategy is financially damaging. If on the other hand, the winery seeks to maintain its profit margin (and if costs do not change within the range of output considered), then the retail price must be set at $8.77 rather than $6.99 when output is increased to 1,500 cases. If production were decreased to 500 cases per month because of a shortage, the gross revenue strategy would indicate retail pricing at $19.99 per bottle, but the gross profit strategy would dictate a price of $13.70. The gross profit strategy tends to restrict the range of price changes. The market situation may prevent a winery from making such price adjustments independently, but at least the profit calculations allow the winery to better understand the costs of maintaining its market position.

Markets are distinguished by a number of factors, including consumer characteristics, product preferences, geographic location, and types of outlets. Within each identified market, however, the competition may be different. That is, individual brands may vary in their market presence and appeal, pricing strategies may be aggressive or passive, and sales may be expanding or dropping. This situation within a market will help determine the pricing point at which a winery's product might be sold. Thus the firm must know how its own brand is perceived by the trade and by consumers relative to other brands at the same pricing point. The brand must be seen as delivering a value that is at least as good as that provided by competitors, and hopefully better. For the salesperson, this comes back to knowing these values at all levels in the distribution and consumption chain.

Pricing practices may differ from market to market according to local regulations or conventions. It is vital to successful planning that wineries know these regulations and conventions and how they affect pricing and other strategies. This is particularly a problem in the United States, where there are so many different regulatory jurisdictions for wine. The most prominent example is in the many different state taxes on wine as an alcoholic beverage. For example, on a wine with an FOB price of $56, Florida charges $5.36 per case, Georgia $3.60, and North Carolina $1.89. Other taxes may tend to discriminate according to the source of the wine. Import taxes are the most important example of such discrimination in areas (such as the United States and the European Union) that are both wine producers and wine importers. Other examples deserve attention as well. In Oklahoma, wine destined for on-premise consumption is taxed at a rate of $1.50 per container, regardless of size. So if a winery is going into an on-premise market with a case of 12 bottles that hold 750 milliliters each, the total tax will be $18. However, a competitor serving the same market with a case of six 1.5-liter bottles will pay $9, or with a bag in the box of 18 liters will pay $1.50. The winery needs to determine if buyers can be influenced to pay extra to obtain the product or if it makes economic sense to absorb the extra cost. Some markets, like Tennessee, charge a 5 percent city tax for products sold to retail businesses but no tax on those sold to restaurants. If a winery is targeting restaurants, it should not factor in the 5 percent.

Some markets are "post-off" markets in which winery prices must be posted with the state and these prices must apply to every customer regardless of quantity purchased. Other markets are quantity discount markets where prices are discounted by 5 percent for 2 cases, 10 percent for 10 cases, and so on. This results in very different pricing strategies for a winery serving such a market as compared to a winery serving other markets.

Pricing decisions involve both the everyday price and the feature price. It is insufficient just to establish a retail pricing point of $9.99, for example. Feature prices are needed to call attention to the product, to solicit sales promotions from retailers, and to motivate distributors. What the winery should remember is that everybody wants a deal—some sort of concession. How much of a deal is required and what proportion of volume it will affect can best be determined in consultations with distributors and others very familiar with the market. Based on these discussions, profit calculations can be made to account for that part sold under a feature price and that part sold at the everyday price. These prices may need adjustment, if possible in that market, to maintain average profits at the desired level. Distributors can be good advisors on what prices are feasible and what is needed to solicit business in any given market.

As the foregoing discussion illustrates, it is difficult to establish a uniform retail pricing point across several markets. Differences in local and regional freight, taxes, and pricing practices lead to different retail prices, even if the FOB price remains the same. If a winery focusing on the southeastern U.S. market has targeted a $9.99 retail pricing point based on its FOB of $56 per case, it will find that it can support that price in only two out of five state markets. Adjusting the FOB price is possible but difficult because it involves new filings with the affected states and complete justification based on defined cost differences. That is a lot of paperwork, and the timing can be critical. Some states do not allow a change in FOB in the middle of a vintage or outside of a defined pricing cycle, for example. In this situation, the use of depletion or special purchase may be appropriate for supporting a targeted retail pricing point.

The nature of the outlets chosen for wine sales will have a significant impact on brand pricing. There is a choice to focus on retail or restaurant chains, on independent retailers and restaurants, or on combinations of these outlets. Chains tend to demand lower prices from wineries, reflecting their buying power, and they insist on uniform pricing across their markets, with allowances for different taxing systems. The costs to the winery of serving the chain's different locations may vary and therefore impact winery profits differently. Pricing strategies in selling to independents or smaller chains may more nearly reflect local conditions, including the relevant winery costs. Prices may be higher than in chain sales. However, the large chain may provide greater volume and economies of scale than independents. Pricing strategies will differ from outlet to outlet, and profit calculations will be needed for each.

While retail chains receive much attention, large restaurant chains have become more significant in wine marketing. Red Lobster is an example. If a winery wants to be with that chain, it will need to consider having the same price everywhere across the country. For regional chains and some independent restaurateurs, it may be possible to establish a price point, for example, in the state of Texas, where a buyer has two or three outlets. My own experience provides a classic example. Our winery was doing programs with a large steakhouse in Louisiana where we needed to be concerned only with prices for that state. The program was so successful that the steakhouse decided to take it national. That was a bit of a nightmare for the winery because some of the pricing could not be applied nationally if the winery was to make any money. Consequently, we had to negotiate a number of price and product issues in order to make the program work within the winery's profit objectives.

WORKING WITH DISTRIBUTORS

Working with distributors is key to successful pricing and programming. The distributor is not an adversary and can be a very useful and needed ally in wine marketing. For this good alliance to exist, however, the distributor must be sold on the product and its potential profit, the winery and its vision, and the winery representatives and their commitment, support, and wisdom. The alliance will not happen unless representatives know what is needed to make the relationship profitable for both the winery and the distributor.

Distributors must earn a margin adequate to cover needed returns on investments in people, trucks, insurance, and all of the facilities necessary to make the business successful. They pay state and local taxes and freight, which become part of distribution costs affecting prices. Therefore, they compare the costs of these items with the level of wine prices and the rate of inventory turn. Over time, they have found that they need margins of 30 to 33 percent of their selling price on most items but can work on margins of 22 percent to 24 percent when negotiating "deep deals" for desirable products. Retailers, by contrast, operate on a "normal" margin of about 33 percent, or a markup of 50 percent on cost. Once wineries and distributors understand their shared interests in making an adequate profit, they can develop a much more effective and pleasant business relationship.

The winery and distributor should establish discrete, realizable goals based on sound market knowledge. The goals should guide all activities so that nothing is undertaken without measuring its potential impact on goal achievement. For example, price reduction programs (such as a feature ad, a better floor display, new distribution, or entry into a wine-by-the-glass program) should have specific purposes. If a winery is giving away money through various support and incentive programs, it should make sure that it ultimately earns a greater benefit because of it.

Some examples can illustrate the importance of discrete and realizable goals. A goal to be in every outlet in which Mondavi is sold is not realizable unless output is large enough to support widespread distribution. The result for a small producer would be extremely thin coverage. A vague goal to be "on premise" is difficult to measure. How many on-premise accounts are there in the market? If there are 750 of them in the marketplace, is it reasonable to be in half of them? If only 10 of the on-premise accounts are white-tablecloth restaurants capable of selling $30 bottles of wine, then a realistic goal for a premium producer might be only 10 rather than 375 or 750 accounts. These decisions can be reached only after the market has been adequately researched to provide the needed information.

If the distributor is capable of executing the agreed-upon tasks and the winery is happy with the results, the winery should not try to micromanage the distributor. However, if the distributor is not doing its job, it will be immediately evident in comparing results with the plans and goals. That is the virtue of well-thought-out plans and readily measurable performance indicators. With such plans, the winery will be able to communicate clearly and consistently with its distributors and develop the relationship that is needed for marketing success.

Knowing freight and tax rates and the region's normal pricing policies is essential for minimizing the cost of doing business. A winery manager needs to be selfish here. While the distributor can be helpful, it may not exert much effort to save the winery an extra nickel or dime. That's the winery's job. If the winery assumes that outlets in other states use the same markup as in California, 50 percent in independent retail stores, then it will lose money. North Carolina has a one-third markup, while Nashville and Memphis, Tennessee, have a 40 percent markup. Freight rates are complex and far from intuitive. Knowing the most cost-effective rates will help maintain profit levels. As discussed earlier, tax rates vary and have an important impact on marketing costs.

Finally, a simple but often underappreciated principle is the need to negotiate with distributors and other sellers. The winery and distributor must have an equitable relationship in which both agree to sell the product, to put it in the right place, to monitor the business problems that occur on a day-to-day basis, and to negotiate the marketing strategy on an ongoing basis. Many times the biggest frustrations for wineries and distributors arise from poor communication and a failure to negotiate fully, fairly, and on a timely basis. Fortunately, there are programming tools that help keep a dialogue open and the winery's products moving through the system. Exhibit 13–1 provides an example of working with a distributor to achieve pricing and other goals.

PROGRAMMING TOOLS

Programming is really about scheduling events designed to move product. The programming plan is established to avoid duplication, unnecessary competition, and misunderstanding. Most distributors try to give equal time to the brands they represent by rotating program periods in which one brand gets a priority for sales, floor display, and advertising and promotion activities. Some firms like to work on 30-day cycles, and others prefer 60-day cycles. Wineries must negotiate a program with the distributor, establishing how the distributor will work with the winery's branded products. If the schedule calls for activity in February, the winery should not call in January to complain that nothing has happened. The program-

Exhibit 13–1 A Florida Pricing Example

> To get a retail pricing point of $9.99 per bottle in Florida, we had to offer the distributor a special purchase allowance of $3 per case (a discount of about 5 percent from our list). This lowered his cost so that he and the retailer could make an acceptable margin. This is worth it to us because Florida is a big market. We wanted to move up our case volume by about 75 cases per month and calculated, based on our costs of goods, that we probably could afford another $2 to $3 discount at the higher movement level. Our distributor said that he needed an incentive program to make the higher sales level, and we agreed to a discount of $2 per case in addition to the special purchase allowance (SPA) already in effect. We specified that most of the volume increase should go to restaurants; we wanted 50 placements and offered $25 per placement as long as the goal of an added 75 cases was met. We still look at the winery price, the tax, the freight, and once again, wholesaler's profit. Now the distributor says "Would you be willing to prorate that? What if we only have a 10% increase?" That's our call. Maybe we split it with him 50/50. Maybe we don't want to come up with all that money. Maybe we say with this kind of program, we'll put up a buck if you put up a buck. Finally, we target this for a specific period—60 or 90 days, whatever is necessary. We negotiate. We try to find common ground. But here we have everything coming together. SPA dollars are working for us to get to a solid retail bottle cost, and an incentive program is working to get a specific goal, in this case, the new placement situation. Hopefully, wineries that do this will have a successful program where everybody makes money. At least they'll have a program against which they can measure performance.

ming plan of action will establish what and when activities are to take place so the winery can make a follow-up call and discuss results.

Calendars are a useful programming tool. Sometimes they specify the dollar amounts of special purchase allowances (SPAs) or other discounts to be given on specific dates. Sometimes a calendar is more goal oriented, specifying, for example, that in March and April, when the new wine lists are written, there is going to be a focused effort to get more restaurant and bar premise interest in the brand. The calendar is designed so that winery and distributor are on the same wavelength and follow the same schedule.

Monetary supports such as SPAs are a programming tool. An SPA is a price incentive given for specific purposes; in other words, it encourages the distributor to purchase the product. It can be done, for example, by offering a $10 SPA off of a $65 FOB price if the distributor buys a pallet load of a specific product that the winery wants to move. Sometimes, the allowance may be made if the purchase is the equivalent of one layer, 14 cases, of a pallet. In that case, the offer may be 14 cases for the price of 13, or about $4.64 off the FOB price. The published list price has not changed, and the distributor may keep all or part of the SPA as extra profit. The distributor may agree to pass on part of the savings as an incentive for retailers to buy special quantities of the wine. If the winery wants to move a new vintage release, it might offer a $5 SPA for the first 90 days of that release. The same might apply to moving a batch of wine that may not meet the winery's quality criteria. In that case, the winery is telling the distributor that the price will apply on the product until the inventory is depleted. The advantage of this pricing strategy is that it is associated with a specific program and it is temporary. Its objective is to move product into the system.

Quantity allowances have been popular in pricing strategies, but they have often been thought of as offering free goods. Buy four cases and get the fifth one free! This is illegal in many U.S. jurisdictions because alcohol cannot be given away as part of a purchase arrangement. It is legal, however, to offer a lower per case price on five cases than on four; so the "one case free" idea is really equivalent to a 20 percent price reduction on five cases.

Depletion allowances (DAs) are intended to drive product through the system, that is, to get it off the distributor's floor. As an example, my winery is scheduled to release a new vintage of chardonnay. However, for distributors to buy the new vintage they need to move their remaining inventory of our past vintage. My winery gives a DA to help them move the product through the system before the new vintage appears. DAs can be very helpful because they offer money on inventory already in the distributor's stock and are specific about product, volume, and value. Many distributors like DAs because unlike SPAs, which are published on a calendar, DA programs are flexible and can be used on short notice for markets of opportunity if the winery agrees. Wineries should be cautious in using allowances. If one gives an SPA on the purchase of a product and follows it with a DA the month afterward on the sale of the product, the winery has given the distributor twice as much money as it may have intended. The general rule is not to follow an SPA with a DA unless there are strong arguments for doing so.

Laws are always subject to reinterpretation, so it is difficult to keep up with requirements concerning pricing and other marketing issues. Generally, it appears that if a winery is doing something in good faith, it will be subject to a warning before more serious legal action occurs. If the winery persists in the same action after it has already been warned, chances are the winery will be fined or lose its right to sell. My winery had an account in Florida, for example, and just recently, our distributor failed to deliver samples to the account for a tasting. Our broker in Florida offered to deliver the wine if we would ship it to him. We accepted the offer because the account was important to us. This turns out to be illegal in Florida because it was a direct shipment to an entity other than a distributor. The winery received a warning letter from Florida regulators and responded with a long explanation of the situation, particularly the fact that a Florida broker had requested the shipment. While we were let off, we were warned never to do it again.

Sales incentives are another program tool designed to improve overall volume and profit. These incentives are outside of price discounts and are direct rewards to sales personnel for achieving specified goals. Awards may be fishing trips, beach vacations, or trips to the winery, for example, in return for selling a designated number of cases or for being the highest-volume salesperson in the distributor's organization. Awards may be for sales to new accounts or to on-premise accounts. A distributor may be willing to share the cost of an incentive program with the winery.

CONCLUSIONS

Pricing and programming are interdependent activities that require effective collaboration with distributors. They must be part of an overall marketing plan that reflects a thorough understanding of costs throughout the marketing system and the conditions in each market. The factors influencing pricing include the nature of brand competition, conventional pricing and margin practices, the availability of various discount arrangements, the relationship between everyday prices and feature prices, and the types of outlets used for distribution.

In working with distributors, it is essential to build an open relationship that fosters productive negotiation and effective pricing. A mutually agreed-upon plan, including discrete and realizable goals, is a key element in a strong relationship. It establishes common actions and provides a ready basis for measuring performance. The plan must account for desired prices, inventory turns, distribution costs, and profit margins. With such plans, the winery will be able to communicate clearly and consistently with its distributors and develop the relationship that is needed for marketing success.

CHAPTER 14

Creating Pull Through

J. Patrick Dore

INTRODUCTION

Winery A has done everything. It has produced the best bottle of wine possible. It has hired high-priced graphic design artists and illustrators; the labels look terrific, the glass mold is fantastic, the closure and cork are beautiful, and the graphic designs on the shipping carton are sharp. Everything is in place. Now winery A has got to take the wine to the marketplace. All the fun stops.

There are three realities in the wine industry. First, there are no brand loyalties in wine. That is why 700 California wineries can exist, along with all the others competing in the market. Second, there is a web of regulations governing what wineries do in the U.S. marketplace. The regulations help protect the smaller wineries so that the larger ones do not just run over the smaller wineries. Third, the United States has a three-tier distribution system. The system is like an hourglass: lots of wineries have to squeeze through a few distributors to reach a large number of buyers. The 700 California wineries work with only three major distributors and more than 30,000 licensed resellers. So the difficulty is in squeezing a winery's product through the few distributors in order to get it to the licensees and ultimately to the consumer.

J. Patrick Dore has enjoyed a long career in wine marketing, including creating his own brand, Dore. Pat is vice president of marketing and sales with ASV Wines, where his focus includes domestic and international sales.

WORKING WITH DISTRIBUTORS

Distributors today are not like distributors 10 years ago. They have become facilitators and service givers rather than the independent merchants they once were. This means that the winery must really manage its marketing program on a day-to-day basis. The winery has to go into a market with a plan that covers all the winery's marketing objectives and strategies because a distributor simply does not have the time to do this for the winery. Wineries can work with brokers and agents much as they do with distributors. Brokers and agents are also the winery's business partners, and the winery can offer them the same type of incentives, the same type of programs, and everything else that it offers distributors.

Currently in California, there are three major distributors: Southern Wine and Spirits, representing about 101 California wineries; Young's Market, with 64; and Wine Warehouse, with 52. There are also various regional distributors that usually carry other beverages such as beer, soft drinks, and water, and each regional distributor represents anywhere from 25 to 125 California wineries throughout the state. Beyond that, there are brokers and agents, another layer of distribution. Or a winery can sell directly through vertically integrated sales companies, direct-to-consumer delivery programs, or its retail tasting room. Not many wineries can rely on direct selling for their entire production, although a few do.

My firm subscribes to a monthly report that describes the competitive situation by tracking about 70 percent of the brands produced in California and their winery (FOB) prices. In the chardonnay category alone there are 20 pages covering the prices of approximately 650 brands. Studying the prices from lowest, $19 per case, to highest, $270 per case, one can identify certain popular pricing points where many brands are clustered. This shows that the chardonnay category offers a lot of different pricing points to consider when pricing a product. By comparison, merlot has fewer and more tightly clustered pricing points.

SALES BROCHURES AND SELLING AIDS

There are a number of ways for a winery to get a product moving to a distributor and through to the final buyers. One of them is a sales brochure. It is important to develop such a brochure; it can provide information about the winery, its employees, its vineyards, the soil conditions, and other important details about the wine and the winery's approach to business. The brochure should convey a certain image to the distributor, the retailer, and the consumer. It should be printed on three-hole-punched paper so it can go into a distributor salesperson's sales book and be used on calls to consumers and the trade.

Recently, I started managing one of our subsidiary companies in Sonoma County. It had developed custom labeling and a good niche marketing program. It had great marketing ideas but lacked focus. So we spent three months creating a really tight sales brochure that explained the company, its objectives and marketing methods, and the benefits for interested distributors. There are four-color product photos throughout, and there is an insert at the back for pricing and other information. Thanks to this sales brochure, sales soared by 33 percent in six months. Although the brochure cost $30,000, it will last for five years—a pretty good investment.

Product photos are important. Simple ones that we call "beauty shots" can be inserted into the salesperson's sales book. One of the most beautiful brochures I have ever seen included a pocket with individual photos of the products and a Velcro closure to keep everything in place. This was gorgeous for presentation to restaurants or chains but not helpful to distributor salespeople because they want a three-hole-punched brochure to put into their book. Additionally, it was too bulky for easy use. Material should be designed so that the people selling the product can use it easily. Other materials include product and winery fact sheets that can be inserted in the sales book, newsletters, and copies of awards or wine writer endorsements. The wine industry is a little like the movie industry: somebody has got to decide what is good, and wine writers are like movie critics, providing endorsements. Other printed material for distribution includes shelf talkers, table tents, and food and wine pairings. Wineries can provide statistical data to those who want to chart product popularity or trends.

There are other sales aids and gifts that can be useful. For example, a terrific point-of-sale shelf talker is the dangler. Danglers are snapped into the aluminum price guide on the shelf; when the air conditioner is working or the consumer goes by, the dangler starts moving, saying "buy me, buy me." Wine racks provide a terrific solution for retailers claiming that they do not have room for a wine. Now they can find more space by putting it on wine racks or in wooden bins. Other gifts include chalkboards, umbrellas, dealer loaders, and bottle openers. Gifts can be quite expensive, but they are very effective.

PLANNING AND INCENTIVES

When calling on a potential distributor, wineries need to focus on where they want the brand to be in the market and what they want to achieve. This requires a completed plan with short-term and long-term programming proposals, including how the wineries will support the brand with point-of-sale materials and other items. Wineries need to discuss how much money to commit for this support and how any costs are to be shared. Wineries must develop a clear understanding with distributors about all of these matters if the partnership is to succeed.

A winery should not overlook the competitive activities within the distributor's organization. It should study the distributor's product book and determine which brands are getting the most attention and why. Is it because of incentives, quantitative discounting, or frequency of programming? This analysis will help in understanding internal competition and identifying appropriate responses, which are both essential steps to take before implementing outside programs.

The challenge to the winery is to create programs that pull products through the distributor rather than push them through. Many elements of a long-term program can help create pull through. The most obvious of them concerns cash incentives. Wineries should know that some incentives are better for the winery than others. For instance, it is better to pay on placements with specific customers than it is to pay on volume. All too often a naive supplier will offer distributor salespeople $10 for every case of chardonnay that they sell out of current inventory. What happens? The salespeople make a single call and sell out the entire inventory and the contest is over. They have earned all their cash with one call, but the winery has not strengthened its sales position. It is better to pay for performance—placements with specific customers or classes of customers. This, after all, is what the winery seeks.

Wineries with a new brand that they want to pull through the distributor should try a high-paying sales incentive program with a definite term, say 12 months, and monthly progress reports. Wineries would have a big board in the sales room listing all the salespeople and their monthly progress toward the marketing goal. That way everyone would know how everyone else is doing. This approach will be most likely to succeed when there is a really super sales incentive. I have used various incentives, including individual and team trips, product or other merchandise, commemorative bottlings, points earned toward buying attractive products chosen from a catalog, and games like superbowl or bingo, where salespeople can earn lottery tickets and other rewards. Other possible incentives are gas credit cards, an American Express card for one year, or limousines to take the salespeople out to a fancy meal at a special restaurant. One of my programs offered an American gold eagle coin—one troy ounce of gold—for the best performance of an individual salesperson, with goals such as placements or adjacencies (getting a better shelf position).

Distributors bill wineries based on the distributors' depletion reports within 45 days after each month. If there are extra incentives, wineries can extrapolate the performance from the depletion report and pay accordingly. Wineries should pay promptly; if they are slow, the distributor's performance will slow down too.

Depending on market regulations, it is perfectly legal to discount wines to certain customer classes and not others. For example, to push a restaurant placement program, a winery could take its 25-case discounted price program and drop it down to three cases for restaurant buyers. The distributor's computer will make the discount on the invoice when it recognizes a three-case order from an identified restaurant. But it will not discount orders to retailers, except on 25 cases, because it has not been programmed to do so. The restaurant price deal is not available or visible to the retail trade because it does not show up in a price book or other document.

PLACEMENTS AND ADJACENCIES

A winery needs to establish clearly understood and measurable market objectives. This helps in directing distributor salespeople to exactly what the winery wants done with its brand. The winery should find out as much as possible about the market (for example, how many licensees are in the distributor's market). Typically, the number could be between 1,000 and 3,000. Imagine that there are 3,000 licensees and a winery and its distributor believe it is reasonable to achieve distribution through 5 percent of those outlets within six months. The marketing objective then becomes 150 placements in six months, and the winery can specify the types or classes of accounts that it wants. This gives a definite target for the distributor. It indicates clearly what the winery wants for its brand.

In addition to placements with target accounts there is the question of where the product is located on the retailer shelf. If the product is on the bottom shelf and customers need knee pads to get it, the product will not sell. The winery needs to specify where it wants its products to be. The best position might be next to the brand leader in the target category. The winery should tell the distributor, for example, that it wants to be next to Roundhill because that brand is flying off the shelf and the winery wants to play off against it. That is the objective of adjacency.

The winery can also tie in its adjacency objectives with case displays and cold box placement. Case displays are a way of showing the product to advantage. Therefore, it is worthwhile to develop objectives such as how many cases should be on display and where they ought to be. The same applies to cold box placement. A huge quantity of chardonnay is sold out of a cold box because people want to consume it immediately. It is a good placement spot and should be included in objectives to guide sales efforts.

Another challenge is for a winery to extend the line of products carried by a distributor. If a winery has been successful with chardonnay, how should it extend its distributor's efforts to sell the winery's cabernet or merlot? The principle remains the same. The winery should structure an incentive program focused on its objective—in this case, line extension. The winery can offer different discounts based on the type of wine purchased: no discount on chardonnay, 5 percent on merlot, and 10 percent on cabernet, for example. In this way a winery can tie multiple products into its discount structure. Or the winery can use an incentive program for the salespeople. One scheme would offer $25 per case if all four in a four-case placement are the targeted variety, or only $2 per case if just one is of that variety. It is legal to do this, and such a program is easily directed.

Most of this chapter's examples have concerned retailers. But restaurants also can be excellent outlets and should be targeted for placements. A winery should give its distributor objectives for placements, participation in wine-by-the-glass programs, selection as house wine, and other features. The winery should give the distributor specific goals that the winery is willing to pay for. I use depletion allowances (DAs) for this purpose and never special purchase allowances (SPAs) because SPAs cannot control what the distributor does with the product as readily as DAs. DAs require the distributor to bill the winery for what the distributor has done against the winery's objectives; the winery, therefore, knows about the distributor's performance.

DISCOUNTS, COUPONS, AND CROSS-MERCHANDISING

The previous discussions have examined ways of providing incentives to distributors through DAs. But a word of warning is needed about price discount programs. Wineries must protect the retail price that they want. They should do everything else to get the product to market, but they should not discount the retail price because doing that is very dangerous. The winery can fall into a pricing pattern that it does not want to be in. A winery should always make sure that it has ample program dollars to get its product to the marketplace. Unfortunately, many proposals I see do not leave any margin when it comes to the buyer unless the retail price target is changed. My arguments with the salespeople center on their belief that the program will work; my experience is that this kind of program will fail because nothing will be left for other program needs. They have blown their entire marketing budget. The answers are to come in at a higher price or to develop some different programs.

Couponing is another approach to selectively discounting the retail price. It works because it is a commitment to the retailer to invest in pulling the product through the store. Our experience and research suggests interesting differences in the use of coupons. Basically there are three types of coupons. One is the mail-in coupon; out of 100 people buying the product, on average only 4 to 7 will send in their coupon and get the discount. Another is the instant redemption coupon in the store, a coupon strategy used by many food manufacturers. Even with this simplicity, it seems that, incredibly, only 4 percent to 30 percent of the coupon holders redeem their coupons. A third is the store coupon, which is part of the store coupon book; the redemption rate for these is 5 percent to 22 percent. Outside of coupons, there are straight-dollar refunds that retailers and consumers like because they do not have to hassle with coupons.

There are cross-over refunds within the category (for example, a discount on a bottle of tonic would be given for purchasing a bottle of gin), and there are cross-over refunds out of the category (for example, a discount on a T-bone steak would be given for purchasing a bottle of cabernet). Cross-over refunds are very popular because they can double the retailer's gross sale. The customer ends up buying both the wine and the steak and not just one or the other.

This type of cross-merchandising can involve a floor display outside the wine area—next to the steaks, for instance. The meat department is one of the longest pause areas in the store, followed in order by the dairy, produce, and floral departments. These are areas where large numbers of customers pause to consider important shopping choices. To take advantage of that long consumer pause, even a chardonnay display could be placed in the meat department. A five-case chardonnay display would also work well in the floral or delicatessen sections, with a coupon tie to those products. It is great to get exposure outside the wine area, particularly if the wine is the only one nearby.

NATIONAL AND REGIONAL DISTRIBUTION

A winery does not have to be national. It can focus on those geographic markets where it can do best and forget the others. I deal with national accounts and do a lot of the pricing calculations discussed in Chapter 13. My software program can display any state I want, with everything pre-priced, the freight calculated, the taxes added in, and the margins already done. But small operators and low-volume producers can select individual states for attention. For example, Colorado is a great state for business because there are no large grocery chains to deal with. Wineries can focus on states where sales costs are lower because there are not the competitive pressures for incentive programs and the like. These states may or may not produce the volume needed to justify the effort, but they are worth investigating. But everyone has the mindset that a winery has to be national. It does not. A winery can select certain states and work them very effectively. Doing this allows the winery to focus better as well as cut down on travel and other expenses. The same principle holds in other national markets throughout the world. A strategy of focusing on carefully selected regional markets will pay off very well for many wine producers.

CONCLUSIONS

This chapter has explored how to pull product through the distribution system. It has recommended actions to take and advised against other strategies. The key active strategies are to develop a clear set of marketing objectives, in conjunction with distributors where appropriate; to prepare a sound, well-understood, and readily measurable marketing plan for achieving those objectives; and to select a set of incentives that will motivate distributors to help achieve those objectives. Objectives include product placement and volume, adjacencies, and line extensions. A sales brochure that is easily used by salespeople will significantly help in achieving marketing objectives, and other materials such as photos, table tents, award reproductions, and shelf talkers are useful.

Performance incentives can be built around DAs that reward the distributors for their performance. Incentives can also include rewards to salespeople, including trips, merchandise, and other recognition. Coupons, dollar-off programs, and cross-merchandising programs are often effective in pulling product through the retail level. Within restaurants this might involve special wine-by-the-glass programs or offers for house wines. The point is for a winery to be creative in offering incentives that do not undermine its target retail price and that do not exhaust its marketing budget.

The list of what to avoid is simple. First, a winery should not use SPAs because they do not allow the winery the control on performance provided by DAs. Second, a winery should not discount its target retail price. The winery should design its programs to support that price or change the price. Third, the winery should not go into the marketplace unless it knows what it wants to achieve, or it will surely get blindsided.

Finally, wineries should recognize that marketing involves a large number of activities, including promotion, advertising, direct selling, product development, and production standards. Whatever a winery does to help pull product through the distribution chain, the winery should make sure it is well coordinated with these other activities.

CHAPTER 15

Advertising

Leslie Litwak

INTRODUCTION

New wine brands are appearing with regularity and making it more and more difficult for any one brand to stand out. That is where advertising and other promotional activities can be effective. They provide a way for one winery to differentiate its brands from thousands of others available in the global marketplace. However, in order to use advertising effectively, a winery must know how advertising is supposed to work.

Two important concepts are commonly used to explain how advertising works. One is a model of how people respond to advertising messages, and the other is a model of how markets respond. The first, if you will, is a model of human behavior, and the second is a model of market behavior. These concepts are discussed in the following two sections. The purpose of the discussion is to increase management awareness of the difficulty of credible market analysis rather than to teach specific quantitative techniques. The discussion describes an organized way of thinking that is critical to management decision making.

Leslie Litwak holds an MBA from Columbia University and has worked in and outside the wine industry with such firms as Del Monte Foods, Memorex, Chateau St. Jean, Sonoma-Cutrer Vineyards, and Paul Masson. She is currently vice president, marketing at the Seagram Chateau & Estates Wine Company.

The Model of Human Behavior

The human behavior model shows that purchase decisions are dynamic, with the buyer moving through stages in which different types of information influence progress towards a purchase decision (Figure 15–1). Advertising first informs people to make them aware of products, then persuades people so that they develop positive beliefs about the products and their benefits. At the next stage, advertising triggers their intention to buy and finally reminds them again to buy, tells them how smart they were to buy, and suggests that they buy again. This model is obviously highly simplified, but it is useful in discussing alternative strategies of market promotion organizations.

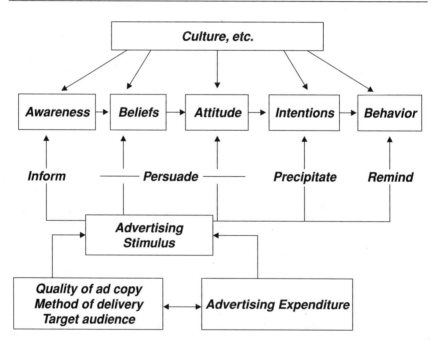

Figure 15–1 A Conceptual Relationship between Advertising and Consumer Behavior.

Advertisers have used this simple model in determining the effect of their advertising message. They have focused mostly on the variables of awareness, beliefs, attitude, and intentions, using such techniques as

- pretests to determine if the messages will reach the designated target
- surveys to determine the number of people that viewed the message, the number of people who can recall the message, and the distribution of positive and negative reactions to the message
- consumer and distributor panels in markets with a different advertising campaign, used to measure variations in buying intentions, brand preferences, images, recall, and attitudes with respect to brands or trademarks

However, it is necessary to have other studies to judge the purchasing behavior of consumers in relation to the advertising message. Advertisers have had relatively little success in showing a positive causal link between the variables cited, awareness and attitude, for example, and purchase behavior. If causal relationships can be estimated, then observed changes in these variables can be used to predict purchase behavior and provide evidence of advertising success or justification for a new campaign. Determining such relationships is very complex and remains outside the realm of much market research. However, econometrics has been used to do this. Econometrics is a statistically based measurement technique used in economic research. It provides a way to analyze such complex relationships. It has been used to evaluate the direct relationship between advertising expenditure and purchase behavior in several important commodity sectors.

The Model of Market Behavior

The model of market behavior is based on familiar economic concepts and is commonly used in econometric studies (Figure 15–2). It is used to determine the consequences of production, pricing, and promotion decisions by suppliers and of purchase decisions by consumers. This model provides the basis for measuring the direct link between advertising expenditures and consumer purchase behavior.

The model demonstrates that if advertising is to increase income to producers, then the demand curve for the product must be shifted out so that

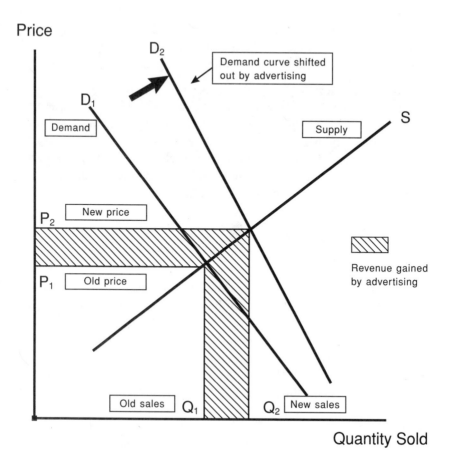

Figure 15–2 A Demand Shift Caused by Advertising.

consumers will purchase more product at any given price. It also demonstrates that if advertising succeeds in differentiating a product, the demand
curve might rotate so that price increases have relatively little effect on the
amount purchased. If the supply is relatively stable, as it might be for
some fine wines, then the impact of advertising is to raise the price. If the
supply of a commodity is unlimited, as it might be for standard wines, then
the effect of advertising is to increase the volume sold. Clearly, most market situations are between the two extremes cited, and advertising has both
price and volume effects. Any analysis of advertising effectiveness must
identify and quantify both the price and quantity changes.

Advertising is only one factor acting to shift the demand curve, and it must be isolated from the other factors. This can be done if the model has been correctly specified, adequate data are available, and appropriate analytical techniques are used. Other demand shifters in the model include consumer income, the price of substitute products, competitive or complementary advertising, and various demographic features such as age, ethnic background, and profession. Government policy is a part of the supply function or the demand function, depending on policy characteristics. Unless markets are analyzed carefully, there is a real possibility of making a wrong decision about advertising.

Empirical analysis using the market model can quantify some important relationships critical for measuring the effectiveness of advertising and facilitating better projections of future results. Most commonly, these relationships are between changes in price, income, substitute prices, advertising, and quantity demanded. They are measured in terms of elasticities, which indicate by how much the quantity demanded will change given a specified change in advertising, price, or other variable of interest. The model will also provide information on marginal costs and revenues that are important in estimating profitable levels of advertising, for example.

Economic theory indicates that advertising expenditures should be increased until the marginal revenue gained from the last case of wine sold just equals the marginal cost of producing that case—including the marginal cost of the advertising program (remember that marginal cost includes "normal" profit). At this point, there is no profit incentive to expand advertising further. Thus, for example, if the marginal return to advertising is greater than one (that is, it more than covers the cost of the advertising), then theory would argue that advertising expenditures should be increased until the marginal return falls to one.

Another way to calculate the optimal budget is to estimate values for the advertising elasticity of demand for the firm's product, the likely response of competitors to the firm's action, and the cross-elasticity of competitor advertising. Economic studies show that the optimal budget is positively related to sales revenue and the absolute value of the advertising elasticity of demand. It is negatively related to the competitor response function and the cross-elasticity of competitor advertising.

A simpler model that ignores competitor responses (as in a monopoly situation) shows that if profits are to be maximized by the marketing entity, the ratio of advertising expenditures to sales must equal the ratio of the advertising elasticity of demand to the price elasticity of demand. This relationship indicates that the proportion spent for advertising should increase as demand becomes more price inelastic. This is the situation for many consumer goods where image is important. In such cases, demand is less sensitive to price, and the advertising sales ratio is high. If the advertising elasticity of demand is low, as it is for some food and beverage products, then the optimal advertising budget is low also. While this formulation does not explicitly account for competitor responses and the effectiveness of their advertising, adjustments could be made to the model to reflect them. This would lower the optimal advertising budget.

If a firm or producer group does not have monopoly power, more complicated models that incorporate competitor response (or other supply response) should be used. The simple model is used here to illustrate some important basic relationships. It shows that there is a sound theoretical basis for measuring the advertising elasticity of demand and determining optimal advertising budgets. Empirical results show that these models have been useful in several commodity sectors.

There is ample evidence that advertising can help increase distribution. It does so by motivating favorable consumer behavior toward the advertiser's product. It provides a tool to help distributors in their sales efforts on behalf of the brand. But clearly, before undertaking advertising, a firm needs to answer the fundamental questions: what is its share of market, where is it selling, can sales be expanded, and does the firm want to expand sales?

It is critical to understand that advertising is just one tool for promoting a product or idea. It is closely associated with promotion, merchandising, and public relations and often under the same corporate management. As an important component of the firm's marketing strategy, advertising should be specified within the market plan. Beyond that, the advertising program should reflect the broader context of other firm strategies, the situation of the industry, and the condition of target markets. Because of the close relationship between advertising and other marketing functions, many of the concepts discussed here are also presented in other chapters of this book, although from different perspectives. This duplication provides essential common threads for building successful marketing strategies.

Each brand needs a business plan that is based on sound advertising principles, clearly establishes objectives and goals, and effectively integrates promotion and advertising programs. The objectives may identify various sales, market share, and profit targets; the creation of product images; and the positioning of the brand against competitors. The goals relate to the timing of strategies and the tactics needed to achieve objectives. The plan should specify the potential advertising budget and a media plan that includes the creative message and how it will be delivered. The media plan should incorporate ideas about using advertising in different ways to presell it to others in the marketing chain and take full advantage of the message that the winery is trying to get across. This chapter considers all these matters.

QUANTITATIVE OBJECTIVES

At the beginning of the planning cycle, the firm needs to decide what it wants the brand to achieve over the plan period. The desired objectives can be defined in both quantitative and qualitative terms, such as the total number of cases to be sold in a specified market, and the target market share to be gained.

The first quantitative decision relates to growth objectives. The objectives must be feasible given the firm's production capability and market category, and the situation in its target markets. If the firm is seeking to extend its line beyond chardonnay, for example, then it will need to analyze the volume and quality it can deliver in other varieties and their likely acceptance in the market. The very basic question is: Does the firm have the capability to grow, or must it focus on maintaining current case volumes? Our winery had some interesting situations in California when low grape production forced a reduction in output and we had to develop appropriate "downsizing" plans. Fortunately, that period has passed.

Market share objectives depend on how the firm defines the competitive cluster within which it operates. For example, a share of the total table wine market does not mean much for most producers because it is a tiny piece of a very large pie that includes giant as well as boutique producers of an enormous range of table wines. A more operational concept defines the market in terms of type, quality, or price and then calculates the appropriate producer share. A market share objective within a specific market can be defined in terms of a varietal category, for example, and then within that, a specific quality and price range. Based on past programs and images established with buyers, the winery may have a brand that can expand chardonnay sales in the $10 to $12 per bottle range but not in other ranges or varieties. In that case, the firm needs to define this price range accurately and discover the total volume moved in that price range. It should assess whether that market segment is growing, declining, or stable and determine if the firm can outperform the market through advertising or other aggressive consumer marketing programs.

Brand managers also need to define market coverage objectives. This is a quantitative measure that indicates where the brand is distributed geographically and institutionally. Expanding distribution may involve either (1) taking a product into geographic markets where distribution is poor or nonexistent, or (2) exploiting new outlets, for example, restaurants or discounters.

QUALITATIVE OBJECTIVES

Creative messages in advertising clearly address many qualitative objectives such as image, awareness, and positioning. A winery needs to make a critical and objective analysis of its current brand image before initiating further actions. One way to do this analysis is to talk to retailers, restaurateurs, and distributors. A distributor in any market sells a multitude of other brands and has the hands-on experience to recognize what the firm's brand image is in the market. Winery representatives should talk to consumers in the winery tasting room, for example, and ask their opinions about taste, value, perceived quality, and what they buy when they do not buy the firm's product. Other ways of gauging brand image are to make surveys, use focus groups, or observe behavior in stores. The purpose is to get a firm fix on what the actual brand image is and what, if anything, should be done with it.

The second qualitative factor is awareness. A favorable image is not worth much if only a few potential customers are aware of it. A firm needs to determine how many people in the target audience are aware of the brand—what it looks like, what it tastes like, what it sells for, and what messages it conveys. If awareness of the firm's brand is very low and that for competitive brands is extremely high because of better promotion or market history, then the proper marketing strategy will be different than if awareness for all brands is low. One of the things that advertising can do, perhaps more effectively than any other element of the marketing mix, is generate increased brand awareness. The firm needs to make sure that the consumer is aware of the brand when making a purchase decision.

The third major qualitative objective is brand positioning and repositioning. This objective is discussed in detail in Part IV. If a firm decides to position a new brand or reposition an older brand, then it will need to develop an appropriate marketing strategy, including a program for advertising and promotion. For example, the firm may have a product that has been very successful at $6.99 per bottle; every available case sells. But perhaps the firm could make more money by elevating the perception of that brand in the mind of consumers and the trade. This would facilitate a price increase by repositioning the brand so that the consumer is willing to pay $7.99 or $8.99 for a brand previously priced at $6.99. The advertising objective is to help the consumer reposition the brand in his or her own mind by emphasizing the special features and benefits of the brand, features and benefits that make the brand more valuable. The message may talk about the link with European fine wines, how great the wine maker is, the vineyards from which the grapes come, or the quality of the wine, from vine all the way to the finished bottle and the label. The messages need to justify to the consumer the repositioning of that brand.

THE ADVERTISING MESSAGE

If the firm chooses to advertise, then it must decide what to say. The message depends on the objectives of the advertising, the product's features and benefits, and the nature of the target audience. The objectives, as noted above, may be to introduce a new product into specified markets, to extend a line of existing products by adding a new varietal wine or new packaging, to reposition a product, or to reinforce a product's current image and position. The situation in the first case is pretty clear cut. Consumers are unaware of a new product and must be informed about it. The message is tailored to identifying the new product and making its features and benefits known to targeted buyers. The message will need to position the product in the consumers' minds, among all the other products available. The message concerning a product line extension or new packaging is different. It can connect back with consumer awareness of the brand name and to related products in the line. The objective is to convince the consumer that the same favorable features connected with the related products will apply to the line extension. In the case of a packaging or labeling change, the message is to alert consumers to look for the change and to recognize its benefits. Finally, messages directed at reinforcing existing product images and positioning can relate back to past positive images, modifying them when necessary to reflect changes in consumer thinking or behavior.

The advertising message also depends on the target audience. There are in fact several audiences that are important to wine sellers. The ultimate consumer may be ordering a bottle off the restaurant wine list, seeking a glass of wine at a bar, selecting several bottles off the supermarket shelf, or buying a case by mail, phone, or the Internet. That is the consumer market. But, in order to get to that market the winery needs to get through the restaurateur, the bartender, the retail buyer, the wine club manager, and the wine distributor. These people are part of "the trade," and a wine seller must get into the mind of the trade if it is to reach the consumer.

Marketing is often viewed in terms of push and pull. Some wine sellers expect that the consumer will pull the wine off the shelf or the wine list. That is, these wine sellers expect consumers to pull the wine through the distribution chain. Advertising is often geared toward achieving that objective. But in order to get the product into a position where the consumer can pull it, the firm first needs to convince the trade to push the product into the system. The trade needs to be presold on the marketing program and particularly the advertising and promotion efforts. The firm should not lose the opportunity to present its program and say, "Look what I've got coming in six months and I want you to be part of it!"

The firm needs to know the demographics of its target audience. Is it talking to an older or younger audience? Is it talking about a wine or label that has a lot of appeal to women, or is it talking about something that is very masculine? Is it talking to a group that buys a lot of wine, or to a group that rarely buys wine? The case of sparkling wine illustrates the importance of demographics. There are a lot of people who buy sparkling wine only if it is their wedding anniversary or New Year's Eve. And there's another segment of people who love sparkling wine and buy it all the time. Advertising in this sector must be tailored to one of these distinct groups because it is hard to get one message across to both audiences.

THE BUDGET

Once the firm has examined its overall situation and decided that the image, positioning, and other objectives can be achieved, at least partly through an advertising message, it needs to determine a budget. This is where context comes into play. There are a lot of different categories in which the firm can spend money. So it needs to determine the financial split between advertising, merchandising, public relations, and sales and trade incentives.

Another factor in determining the budget is the competitive situation. The firm should learn what competitors are doing. Are they advertising or not? How much are they spending? How are they doing it? What is their message? How effective are their efforts? A high level of competitive advertising may require a larger-than-planned advertising budget if the program is to be successful. Or it may suggest a greater emphasis on merchandising or public relations. A low level of competitive advertising may make the firm's efforts more effective. In either case, knowledge of competitor actions will permit the firm to speak intelligently to distributors, saying, for example, "We know Brands A, B, and C are doing this. But we're bringing you a unique program that will set ourselves apart and make your job easier." Or maybe A, B, and C are not doing any advertising, so the firm can say, "We are the one in this competitive set that's going to support your efforts with consumer advertising and an effective message."

There is the question of national as opposed to local or regional advertising. Can the firm afford to advertise in New Jersey, New York, Florida, and California? If it cannot, then it must review advertising response calculations and decide in which of the states it should advertise. Or it might consider altering the frequency of advertising messages or the media chosen to deliver them. Clearly, running an ad in national print or national TV is a lot more expensive than going into particular geographic regions and buying spot radio or spot TV. The various media include TV, radio, outdoor billboards, and print, either general or specialized. Direct mail and the Internet are very quickly becoming a logical and potentially effective spot for advertising. The decisions about message and media relate to the firm's assessment of its current distribution strengths and what its distribution objectives are.

The advertising budget is split between the cost of producing the ads and the cost of delivering them. When most of the budget is spent in the creation of the advertising, then relatively little can be spent on its dissemination. There are some notable instances when the ads were so creative or clever that they achieved their objective even though the frequency of presentation was low. For the most part, however, more money should be allocated to delivery than to creation.

SCHEDULING

There is a big step between planning and implementation. The planning process must allow ample time to put the program into place. An advertising plan put together in July for use in October is unlikely to produce its full benefits because the time allowed for implementation and for consumer behavior responses may be inadequate. Seasonality is an important scheduling consideration. Would a producer want to advertise sparkling wine outside the October-November-December time period? Probably not unless it had a very, very large budget, because that period is when most people buy sparkling wine. The last quarter of the year is a peak sales period for many different wines and is a logical time for most producers to advertise. Given this concentrated period of competitive advertising, it may make sense to consider counterseasonality: either to focus on those clients who buy sparkling wine throughout the year, or to go in with a program for summer, when no one else is advertising and when sparkling wine might have appeal as a cooling drink. This strategy could provide a greater return on the advertising dollar. This might provide an incentive for the distributor and the retailer to put the product on the floor where people can see the product, learn about it, and buy it easily. They will have read or heard the advertising and be more likely to buy.

The schedule for advertising is integrated with that for other elements of the marketing strategy into a calendar called the merchandising advertising calendar (MAC). The MAC provides a ready reference for the context within which the advertising program will be implemented. Managers can see how advertising plays with the rest of the marketing mix and determine if the firm is communicating a logical and consistent message across advertising, public relations, and merchandising.

As an example, a calendar may show the following events. A new vintage chardonnay is going to be released in April. In January, the firm will send press releases announcing the new product release. This will be followed by sending out samples to the same audience in February. Great! This is the audience that is going to taste, review, and talk about the wine. During April and May, the firm will have a promotional program giving the trade a little larger discount to stimulate shipments into and through the system. It may also offer depletion allowances to keep the product moving out from the distributors. During this period, the firm will send its wine makers to do a series of wine maker dinners that focus on the new release. In May and June it will be time for merchandising at the retail level. This may involve a case card and a necker for the supermarket, or a little table tent promoting this wine by the glass in a restaurant. At the same time, the firm will advertise so that the consumer becomes aware of the new product. So, the consumer sees or hears the advertising, and then walks into the store or restaurant, sees the merchandising material, and receives a further prompt to purchase the product. It has all come together.

Notice the sequence in which the various promotional activities are scheduled. Press releases are used before the product is available, then samples are sent and wine maker contacts are arranged; this is followed by incentive and merchandising programs; and finally all is coordinated with an advertising campaign. Advertising was not used before it could be effective. Merchandising materials were made available first with distributors and then retailers to stimulate the flow of product into and out of the warehouse and retail outlets. Advertising was used to provide consumers with information and to persuade them to search for the wine. This is the logical sequence that results in profitable product sales.

PRESELLING THE ADVERTISING CAMPAIGN

Once the advertising and media plans have been developed, the advertising needs to be sold. The firm should make sure that every audience along the way, including its own sales force, the distributor network, and the retail distribution system, is aware of the campaign and its objectives and benefits. This will provide a greater return for each advertising dollar than can be earned from a campaign that is not presold.

One effective presell strategy is to replicate the advertising; for example, a firm can reproduce a print ad on a sell sheet used with distributors. The sell sheet will then show what the ad looks like, where it is going to run, and when it will appear. This can be used by the firm's sales force and by distributors and retailers to gauge the support they will receive and plan their own selling efforts. They become part of a coordinated sales team committed to making the campaign work. In effect, the preselling strategy is a means for managing the distributor network by keeping it informed and involved. This provides a competitive advantage to a winery facing distributors with a wide range of brands available to them.

As an example, consider a winery that wants to improve its distribution in a particular market. To do this, it may be important to gain sales in a specific supermarket chain and in the wine-by-the-glass program of some key restaurants. The winery will need to coordinate its merchandising, public relations, and advertising to focus on the target market. It will need to get ideas from the best distributors in the area as a way to improve the campaign and to gain distributor involvement and commitment. It should specify the schedule of sales activities well in advance so that the distributor can coordinate effectively. The winery should specify what it expects from the distributor in exchange for this sales support. In this way, the winery is using its marketing program as a vehicle to get effective distribution in the new market.

Even if its distribution in a specific market is satisfactory, a winery may seek to improve its display and feature activity in the important selling months of November and December. It wants displays on retailer floors and wants to be featured in store ads. It wants to be prominent in wine-by-the-glass programs. These objectives, and the sales support offered to achieve them, must be worked out with the distributors in advance. The distributor will not know what's expected of it unless the winery spells out its objectives, and the supporting activities that will help get the consumer to pull the product off the floor or to order it in a restaurant.

COMMUNICATING DIRECTLY TO KEY MEMBERS OF THE TRADE

One of the primary tasks of the winery sales force is to communicate directly with members of the trade. But this should be supplemented with other communications, for example, letters directly from the winery, news bulletins, and wine maker interviews. Although salespeople can speak directly to customers about the advertising program, the salespeople's impact may be enhanced by something "new," such as a letter from the winery about the upcoming program for the holiday season and the advertising that will accompany it. This is not very costly and allows the winery to communicate directly to those in the trade who ultimately will decide whether to put the brand on feature in December, to recommend it to customers, or to give it preferential shelf space. The objective is to ensure that people in the trade are working on the winery's behalf as well as their own.

CONCLUSIONS

This chapter has described a context for the questions to be asked in developing a marketing plan and determining if advertising is a strategically appropriate course of action. It discusses how to make advertising more effective once it has been produced. There are two key points that deserve emphasis. The first is that advertising is just one element in a total marketing strategy. It can be powerful and effective but needs to be well coordinated with other marketing efforts. Second, the impact of wine advertising can be enhanced by ensuring that other members of the distribution chain understand the message and buy into it so they can help achieve its objective.

CHAPTER 16

Category Management

Alex G. Franco

INTRODUCTION

The range of wines available throughout the world is enormous. In the United States during the mid-1990s, there were more than 3,900 individual wine items stocked in supermarkets and identified by Universal Product Codes, according to WineScan Reports published by the A.C. Nielsen Corporation, a major research organization. And there are many more items stocked elsewhere, for example, direct sale wines or wines without product codes. But Nielsen data indicate that only 340 of the large number of listed items accounted for 80 percent of the dollar volume. The remaining 3,560 items shared the remaining one-fifth of dollar volume. The question facing retailers is obvious: what combination of wine items will deliver consumer value and maximize overall profits for each store? Category management, the process of treating product categories as strategic business units, is a means of answering such a question. In the case of wine, it facilitates identification and handling of the assortment of wine items that will provide a high level of both consumer value and store profits. Category management helps eliminate unnecessary duplication and waste.

Alex G. Franco received his BA in psychology from Rutgers University. He has worked in marketing and category management for UDV Wines and Safeway. He currently is employed as sales operations manager for customer interface sales at The Clorox Sales Co. in Oakland.

Category management is part of a concept called efficient consumer response (ECR). ECR is based on a coordinated two-way information network among producers, distributors, retailers, and consumers that seeks to match buying and selling needs throughout the distribution system, thereby increasing efficiency and delivering better value. Category management has become a way of life in the U.S. distribution system now as more and more producers and retailers have adopted it. But the alcoholic beverage industry has lagged behind other industries in adopting category management.

From a producer's standpoint, category management entails a working relationship with a large retailer in which the producer becomes a partner in managing a specific product category with the retailer. It involves a willingness to protect and enhance the category and goes beyond protecting the producer's own brand. The producer contributes knowledge of the category and its unique characteristics, and the retailer contributes knowledge of category relationships and retail shopping behavior. Retailers most likely can handle category management by themselves, but they can benefit from suppliers' insight on matters with which the retailer may not be familiar.

Category managers for a large chain may be responsible for beer, wine, and spirits, in addition to other categories. So they may not have the time or the expertise to understand the unique characteristics of consumer behavior for just one of the assigned categories. Collaboration can be the answer that allows each of the partners to enhance profits. To be successful, the producer must demonstrate continuing objectivity and credibility. This requires familiarity with the retailer's objectives and strategies, and the ability to balance producer marketing objectives against the "right" category management decisions. This is a dynamic process in which decisions must be monitored and adjusted continually to reach agreed-upon objectives.

CATEGORY DEFINITION

Once the producer and retailer have accepted this management concept, the category must be defined. Broadly speaking, a category contains a distinct, manageable group of products or services that consumers perceive to be interrelated or substitutable in meeting a consumer need. This definition is not as simple for wine as it might seem. The key objective is to determine what the consumer believes can be substituted for the product and what the consumer believes complements the product. The consumer wants substitutes and complements to be near each other. For example, if consumers believe that frozen waffles and sausages are complementary breakfast foods, then they want to find the products close together in the store. This makes it easy to shop the breakfast food category. Depending on consumer perceptions, the wine category can be narrowly limited to just domestic table wines, defined to include all table wines that can be substituted for one another, or expanded to include other wine types (for example, sparkling and dessert wines) as well as other beverages and complementary goods such as wine glasses.

CATEGORY ROLE

The next step is to determine the category type or role that it plays in the retail marketing strategy. Essentially, there are four types of product categories in a supermarket: destination, routine, convenience, and seasonal or occasional. A destination category contains products of sufficient value to consumers to warrant a special effort to acquire them. From the store's perspective, these are the categories that establish the retailer as the dominant store of choice by providing the target consumer with consistent, superior value as measured by assortment, price, and service. Examples include wines and spirits, specialty foods (organic, ethnic, exotic), and kitchenware. About 10 percent of supermarket products are in this category. It is an expensive category to maintain because it requires advertising and promotion, the allocation of significant resources, and continual monitoring and adjustment.

Routine categories account for about three-quarters of all supermarket categories and include products routinely purchased for stock-up needs, such as canned foods, breakfast cereals, and dairy products. For these categories, the store focuses on providing consistent competitive value in meeting consumer needs.

Convenience items, accounting for 10 percent of the products, are managed to reinforce the retailer's image as a full-service store by providing good value to the target consumer in meeting less-planned fill-in needs. Convenience items include motor oil, shoe polish, light hardware, and other nongrocery items that make shopping convenient for customers and keep them in the store.

Seasonal and occasional categories include products that are mostly purchased according to seasonal needs or bought only occasionally throughout the year. Items in these categories include Halloween candy, suntan lotion, and sparkling wine. The categories are managed to reinforce the image of the retailer as the store of choice by providing timely, competitive value to the target consumer. These categories make up 5 percent of all goods in the store. Management can develop strategies to shift products from one category to another, depending on the strength of consumer perceptions about the category. For example, certain convenience items like paper products may be shifted to a destination category, if the consumer believes that there is a full range of qualities that are important enough to warrant a special trip.

Retailers typically address three issues in determining which role is appropriate for a specific category. The first is the relative importance of the category to the consumer. Wine, for example, is purchased by a relatively large portion of households with above-average incomes and is likely to be of importance to such households. If such buyers are important customers for the retailer, then wine should be in the destination category. If, on the other hand, these customers view wine as a routine purchase, then efforts by the retailer to make wine a destination category are unlikely to be worth the expense. The second issue is how important is the category to the retailer. If the category does not contribute significantly to profits, or if it lowers sales in more valuable categories, then it may be maintained only for customer convenience. Finally, the retailer is concerned with the category's market outlook. Wine has had a mixed reception in this regard. European table wines (*vin ordinaire*), U.S. economy wines, and some other types of wines have been in decline, and sales of these types of wines are expected to continue falling. This makes retailers wary of these wines. On the other hand, premium or fine wines have experienced tremendous growth worldwide, and this growth has been attractive to retailers. The degree to which this outlook will influence retailer decisions about the appropriate role for the wine category depends on the first two issues, how important the category is to the consumer and the retailer.

Wine is attractive as a destination category because it tends to attract upper-income customers into the store. A strategy for making wine a successful destination category will have several elements. One element is a balanced assortment of wines, from the economy to the ultrapremium levels, that will offer the opportunity for customers to trade up as their situations change. A second element is shelf organization. To facilitate shopping decisions, various product types should be organized in a price flow arrangement. Wine should be cross-merchandised in multiple locations (meat section, delicatessen shop, pasta aisle). This will allow more frequent exposure to potential buyers and encourage impulse buying. Finally, the shelf displays should be coordinated with cold box displays. The cold box should invite self-service at multiple price points and provide real convenience for the customer.

CATEGORY ASSESSMENT

The next step in the business process is to evaluate the category. This is a process of collaboration among producer, distributor, and retailer because no single agency will have all the data or insights to effectively perform the analysis alone. The task can be divided into three stages. First, there is data acquisition. Second, there is data analysis. Finally, conclusions are drawn and the implications of those conclusions for business decisions are assessed. Category assessment may be the most difficult part of category management, but it is essential to success because it identifies the gap between the current state and the desired state and uncovers key business-building opportunities. The assessment should cover market conditions, the retailer's situation, supplier competition, and consumers' preferences and behavior.

The market assessment seeks to identify the sales and consumption trends of the category, and the nature of the competition. It also should calculate the market share for the category gained by the retailer (or distributor) and competitors. The assessment of the retailer or distributor seeks to discover relevant sales and profit trends, inside and outside of the category. It should also identify and evaluate, if possible, other factors affecting the retailer's market performance. The supplier assessment focuses on the efficiency and profitability of each of the category's suppliers. The objective is to gauge future competitive conditions.

The objectives of the consumer assessment are (1) to identify who buys items in the category and describe their demographic profile, including age, sex, income, lifestyle, and household purchase penetration; (2) to discover why they buy items in this category and what factors are important to their decisions; and (3) to determine the frequency and pattern of purchases and where they are made, and describe how consumers buy within the category (for example, What is the package size? Are there related purchases? What else is in the market basket? Was the shopping trip planned or unplanned? Was the purchase promoted? Are consumers loyal? Are there brand- or store-switching patterns?

A key objective is to understand how the consumer thinks about and makes purchase decisions. It can be helpful to create a decision-making tree that identifies the sequence in which purchase decisions are made. For example, the nature of the use occasion may result in a series of product-related decisions that will change as use changes. The sequence could be as follows: occasion, type of beverage, color, variety or type, price range, size, and brand. Obviously, the decision points that are significant for average wine buyers may not be significant for sophisticated buyers. However, knowing the correct sequence may significantly influence how wines are organized on the retail shelf and presented in promotional material.

The decision tree framework encourages research concerning brand choice and its influence on buyer behavior. Survey results indicate that brand loyalty seems to be inversely related to wine price class. It appears strongest among economy wines and weakest among premium wines. These findings will vary among markets and brands but have significant influence on how a category will be defined and managed.

CATEGORY SCORECARD

Once the assessment has been completed, the retailer and producer need to determine target objectives for the activities described in the business plan. This process can be carried out through the development of a "scorecard" that lists the various criteria against which performance will be measured, and the level of achievement expected within each criterion. Table 16–1 shows part of a scorecard developed by United Distillers and Vintners.

Of course, the business plan may include different activities from those listed. For example, a company might establish objectives on the number of transactions, customer counts, purchase occasions, or the number and frequency of market surveys. The plan may specify the development of a monitoring system that will provide information on transaction values, promotional effectiveness, and market share and profit among collaborating retailers. Of course, any criterion chosen for evaluating results must be

Table 16–1 Category Scorecard

Criterion	Current Status	Objective
Category share		
Of the department	_____	_____
Of the market	_____	_____
Consumer		
Consumer service level	_____	_____
Transaction size	_____	_____
Satisfaction	_____	_____
Sales		
Category value of sales	_____	_____
Growth	_____	_____
Sales per square foot per week	_____	_____
Profit		
Category gross profit	_____	_____
Gross margin	_____	_____
Gross profit per square foot per week	_____	_____
Product supply		
Days of supply	_____	_____
Inventory value	_____	_____
Inventory turn ratio	_____	_____

measurable by a credible and feasible technique. The producer's ultimate objective is to determine if the category management program is profitable.

STRATEGIES, TACTICS, AND IMPLEMENTATION

Category management works only if the collaborators agree to a plan of action that includes strategies, tactics, and implementation. The advantage of category management lies in a planning process that compels managers to think about the category, to test the assumptions about it, to choose specific actions that are most likely to achieve target objectives, and to adhere to a system that measures results in a meaningful way. The choice of strategies is driven by the category role. What is right for a destination category may be too expensive or otherwise inappropriate for a routine or convenience category. For example, the marketing of a destination category tends to be more aggressive to avoid having the category be overtaken by competitors, and advertising may be much more intense. The basket of strategies includes those concerned with marketing, promotion and advertising, financial commitments, and inventory. Strategies define what to do to reach target objectives. Tactics are the specific actions to implement category strategies. They are developed in the following six areas: assortment, pricing, product supply, shelf presentation, merchandising, and promotion. Once this stage in the business process has been reached, implementation of the plan can begin. Implementation leads back to the beginning again, with a continual review of results and adjustment of category definition, role, assessment, scorecard, strategy, and tactics.

GEODEMOGRAPHICS

Geodemographics is one of the most powerful tools available to category managers. It allows them to understand the target market for wine perhaps better than the retailer understands it. It can profile customers, determine product preferences within a geographical area, and identify geographic areas with high concentrations of targeted customers. Combined with historical sales data (Scandata) from the retail chain itself, geodemographic data can provide both horizontal and vertical perspectives on the consumer population and clearly show the strengths and weaknesses of the retailer program. It can identify the chain's customers and what they are buying. It may reveal, for example, that its customers are running contrary to others and do not drink chardonnay, and therefore, that the retailer should cut its chardonnay assortment and add to its red wine inventory.

The first part of the geodemographic-data-gathering process is mapping. Each chain outlet is coded by address or ZIP code and assigned a latitude and longitude by the mapping program. Each location is surrounded by a concentric circle 2.5 miles in diameter, to define a comparable but arbitrary trading area. The trading area is subjective, but the objective is to analyze the people that live around the store to determine where they shop, what they buy besides wine, how much money they make, and what other factors influence their buying behavior. The other variables measured include ethnic mix, lifestyle, education, and age. The demographics concerning income, education, age, and ethnicity are derived from census data that report down to the actual city block on which respondents live.

Other data are from consumer surveys and panels. These data are important because income, lifestyle, age, and educational attainment are important factors influencing wine consumption. People with more education, higher incomes, and certain lifestyles tend to drink more wine than others.

The data must be analyzed and their implications for marketing discovered. The value of the results is enhanced if scanner and other consumer survey data are added to the census data. If the analysis is done properly, it will aid the producer in deciding the best markets for its products, creating an efficient assortment for the retailer, clustering groups of stores by different buyer types, and identifying geographic areas with a large concentration of buyers.

Table 16–2 illustrates what can be learned through geodemographics. It is drawn from a trading area population analysis done for two stores of a local chain and has been simplified to show relatively few variables. A full-scale market study will obtain data on a wide array of variables that experience has shown to be important determinants of consumer buying behavior. In this example, the comparison is between the populations served by store A, located in a densely populated city, and store B, located in an affluent suburb. Population characteristics differed, as might be expected. Once the analysis of all six stores was done, those stores serving similar populations were grouped together under common marketing programs.

Different demographic characteristics lead to different wine-buying propensities. In measurements of these propensities, relative preferences for different wine products emerged (Table 16–3). Customers of store B more strongly preferred table and sparkling wines than did customers of store A. Although the data are not listed below, they revealed that store B customers had stronger preferences for premium wines than did store A

Table 16–2 Geodemographic Comparison of Stores A and B

Characteristic	Store A	Store B
Median household income (U.S. $)	$47,360	$75,546
White	57.7%	80.9%
Asian	34.5%	16.0%
Black	6.0%	1.7%
With bachelor's degree	23.3%	26.8%

Table 16–3 Index of Relative Purchase Propensity by Product for Stores A and B

Product	Store A	Store B
All table wines	96.4	100.0
Sparkling wines	73.0	100.0
Sake	100.0	81.6
Nonalcoholic wines	100.0	95.6
Coolers	100.0	94.9

customers. On the other hand, store A customers had stronger preferences for nontraditional wine products such as sake, coolers, and nonalcoholic wines. The implications for category management decisions are important. For example, they suggest that store B might be able to develop the wine category as a destination category and that store A might consider an emphasis on nontraditional wines to take advantage of customer preferences for those products.

Another example illustrates how geodemographics can be used to test ideas about how buying behavior is influenced by demographic factors. In this case, a major producer used scanner data from 300 stores of a regional chain to identify those stores with relatively high sales of premium wines. It then analyzed the demographics of the trading areas served by those stores using U.S. Census data and found some common factors that separated the high-selling stores' customers from those of the remaining stores. The household members in the trading areas surrounding the stores with relatively high premium wine sales tended to have higher income, more professional jobs, better education, and higher rent. Household income was one-third higher than the U.S. median, and the number of both college graduates and persons in white collar jobs was 14 percent above the U.S. average.

A large producer and supplier of wines may do this sort of analysis for the major stores it serves, covering a wide range of domestic and imported products, including table wine, sparkling wine, coolers, and substitute products. To be credible, the analysis must account for the large number of institutional, economic, and social factors that may be in play in a specific market. In this respect, store-specific scanner data are invaluable when used in combination with broader market data. The objective is to give the retailer as much information about its consumers as possible, thus demonstrating the producer's value as a partner in category management.

In summary, geodemographics is a management tool that will help in selecting the best markets for the product, creating an efficient assortment for the retailer, guiding the grouping of stores by different buyer types, and identifying geographic areas with large concentrations of buyers.

CONCLUSIONS

Personal observation shows that wine is an important category in supermarkets and that sales have grown over the past decade. Some of this growth may have resulted from the creation of supplier-retailer partnerships to improve management of the wine category in retail stores. The success of this effort suggests that the idea should be examined by others and applied to markets where it is not now prevalent. The keys to success include innovative thinking, objectivity, and the use of credible technological tools such as geodemographics, scanner data, and valid analytical frameworks. Finally, effective category management recognizes that consumer decisions drive the process. Through a better understanding of consumer decisions concerning wine purchases, retailers and suppliers can do a better job of profitably serving consumers' needs.

CHAPTER 17

Label and Bottle Design

Ralph Colonna

INTRODUCTION

The purpose of this chapter is to increase marketing managers' understanding of what is involved in label and bottle design. This understanding is critical to managers' decisions about product differentiation strategies. The key message is that proper design leads to effective differentiation and increases sales. Proper design requires recognition of the links between packaging designs and those used in promotion, advertising, and other winery activities. The chapter begins with an outline of the principles involved in label and bottle design, then considers certain aspects of label design, bottle design, and finally, costs and research.

PRINCIPLES

Decisions about label and packaging designs should take into account the price point of the wine. Labels that deliver the wrong message about the wine are ineffective or damaging to sales. Consumers may be deceived one time by a $20 label on a $5 wine, but they are unlikely to buy the wine again and may be hesitant to buy any product of the winery. And one well-known proprietary blend had a label that looked like the label of a $15

Ralph Colonna attended the Art Center College of Design and the Otis Art Institute in Los Angeles. His career in design spans 40 years. For the past 25 years, his firm CF/NAPA has specialized in luxury consumer goods with a particular emphasis in wine marketing.

wine but it actually was trying to sell a $45 wine. The basic principle is that if the consumer is going to believe the message, then the message must be correct.

The graphics used on the label and in packaging and promotion should be fresh and new and should associate the product and the producer with positive values—values that consumers will respond favorably to, such as high quality, enjoyment, a good reputation, and confidence.

A very practical principle relates to mechanical requirements. Both label and bottle design must be compatible with existing machinery or manufacturing processes. If they are not, the benefits achieved by the new design must be sufficient to justify the costs of changing machinery or processes. However, creative uses of machinery, such as altering a sprocket ratio or a die size, might make some very unusual designs feasible, without much change in cost.

Label design should be based on a comprehensive statement of what the winery wants to accomplish. This marketing brief should include what image is to be conveyed, what the wine will taste like, what the price position will be, and which consumers are to be targeted. This represents the platform from which the label design effort is launched.

Label size decisions depend on what information the winery wants to convey and what information the consumer prefers to have. As an example, some wineries prefer rather large labels that tell a story about the wine and the winery. Many consumers, on the other hand, prefer less label area and more glass so that they can see the product inside the bottle. Labels for large bottles (such as 1.5-liter bottles) should be larger or otherwise different from labels on standard-size bottles. Larger bottles have more visual impact if the design of the bottle and the design of the label are well integrated.

Label designs need to be legally defensible. That is, the design should not infringe on an existing trademark, patent, or copyright or violate relevant governmental regulations. Interpretation of what is legally defensible and what is not is quite difficult, as evidenced by the long and expensive legal dispute between Kendall Jackson and Ernest & Julio Gallo over the leaf design used on their labels.

The basic rule on label messages is to talk to the target consumer, not to other wineries or to others who are not target consumers. There is no "best" message for the back label. The text depends on what consumers will react to and what the winery wants to communicate. It may be worthwhile to tell the history of the winery, or the origin of the particular wine, but it is a waste of time to talk about what food matches the wine. Most consumers know this already, or do not care, or will be discouraged because the label indicates the wine is good with shrimp but says nothing about the Dover sole that is on the menu for that night.

Label designs are not forever. In the 1970s, one could expect a label to be effective in the marketplace for 10 years. Since then the time period has been shrinking. For major brands, it is now one to two years. Labels are being continually modified, perhaps not in the major components such as the brand logo, but certainly in other elements. The continuing objective is to make sure that the label and packaging are vibrant and that they work effectively in selling the product.

Bottle styles are not forever, either. The focus is on adapting the style to the needs of the marketplace. There is an emerging trend to diversify bottle shapes. Wineries are using embossed bottles, rounded rather than round bottles, and long and thin or short and squat bottles. There is no reason that domestic wines need to be in Bordeaux- or burgundy-style bottles, or in traditional colors. Why not use blue, red, or yellow bottles if that helps sell the wine? The ultimate criterion is what style or sequence of styles does the best job in the long run to sell wine for the winery.

WINE LABELS

There are different perspectives about what a label should convey. The producer wants to tell the most complete story possible, which usually means between 25 and 30 pages of a book. Now how will that fit on a label? The sales manager wants labels that jump out and say "buy me," in the hope of gaining market share. Regulators have a different vision of what is important and how it should be communicated. The retailer wants, among other things, a label that is easy to scan. The consumer looks for a fair indication of what is in the bottle and what the benefits of purchasing the wine are. Research and market experience have shown that the more complex and finely detailed the label, the more consumers will perceive the product to be complex and finely detailed.

The designer is torn between aesthetics and profit. The design may be something beautiful but impractical to put on the bottle. There are a lot of factors to consider: the type and price of the product, target markets, and consumer demographics. Some design variables are restrictive (such as budget, bottle size, and mandatory statements), and others are flexible (such as label shape and placement; the relationship between labels, bottles, and other packaging; and color and printing options). Designers must decide on, or recommend options for, paper type and quality, the use of die cut, the number and place of borders, the use of calligraphy, and other items. They need to consider the typeface to use on the label, cartons, and foil stamping. They also must know when to stop working on a particular label.

A designer is usually called in when a winery needs a label and wants to market it right away. However, the work often develops into a complete identity package for the winery. This will include a logo, stationery, signs, catalogs, newsletters, and tasting-room designs. The designer is immersed in the winery's special culture for the full period of the project.

A case study illustrates some of the considerations that go into label design. One client established a winery in Northern California in the mid-1970s to produce top-of-the-line varietal wines. Work on label design began well before the product was ready for shipment. My firm produced 30 to 40 idea sketches and developed a proprietary typeface and script that could not be copied legally. We helped develop the winery's logo. One concept of the winery involved emulating French techniques with a California twist that identified the wine as a California product. We took the engraving style of illustration used in Europe and combined it with contemporary (and proprietary) typefaces. The engraving was cut on a very small metal block that was the size of the finished drawing. This was time-consuming work and relatively expensive. But it conveyed the message that the winery wanted. We also developed a slant-pattern design that appeared on the label, the carton, and promotional material. While we have changed the design over time, it is difficult to recognize the differences unless the labels are side by side. One of the keys to changing a brand is to keep it viable without lowering the brand equity.

Another winery plagued by disappointing sales results asked us to redesign its label. We got rid of the old label design, typeface, and logo, and developed a new, cleaner image that really jump-started sales in a short time without a change in product. We went on to do a complete package for the winery: stationery kit, little invitations, brochures, wrapping paper, bags, carry-out boxes, and shipping containers. Part of the effort was to create really positive floor stacking display units that would increase sales. Design efforts like these need to be coordinated so that similar concepts flow through all winery materials.

GLASS CONTAINERS

Designing and testing new glass containers involves several steps. Initially, the designer, working with the marketing department, makes drawings in outline or silhouette form of what the bottle might look like. The winery then tries to determine what is a pleasing shape and what is not. Models of proposed bottles are first produced in solid plastic, such as Lucite, using a computer. The designs can specify various formats, for example, fatter, thinner, taller, or less round. From these, the computer will control a milling machine that produces each model. This allows models to be produced for $200 to $400 each. Using models allows the designer to consult with the winery as new ideas are produced and obtain suggestions for future design changes.

Bottle coatings will be used more and more in the future to differentiate one glass package from another. Another innovation is the crackle finish. The bottle is not fragile and will stand up to shipping and handling. We designed another bottle with the label silk-screened and fired directly on the glass. It had a neck label with the look of gold patina that required six months of printing explorations to develop. The ACL printing process uses ceramic fired right into the glass to make the logo or label. Bottle manufacturers are more willing to consider these processes today than they were in the past. For a client some years earlier, the only thing we could get was a single initial in the glass plus a change to the finish and top part of the bottle. Since that time, of course, Mondavi and others have adopted the flange-top bottle, and most wineries are trying different bottles. One design goal is to incorporate some attributes of the product directly into the appearance of the label and bottle.

The following case illustrates how a bottle can be used in product differentiation strategies. The client, a sparkling wine producer, believed that the bottle style was going to be an important way to distinguish its new line of sparkling wines. It wanted a proprietary shape that would make a big impact on the market. So we designed what we called a "Euro-style" bottle in a very sensuous shape. It was based, for the most part, on what was being done in Europe at the time, and it was in fact made in France. The bottle had a very long, thin neck. Although the neck was criticized by some as restricting wine making and bottling practices, it did not turn out that way. The winery logo was silk-screened onto the glass and fired right into it. There was no way to get it off. The foil over the cork was very short to allow for a neck wrap that went up and under the foil. When the cork and foil were removed, the neck wrap with the winery logo would still be visible, even though the bottle might be in an ice bucket. There are two labels on the bottle: one carries all the mandatory statements and the other has the winery logo and eye-catching design, which is the front label for display purposes.

COSTS AND RESEARCH

Costs vary tremendously depending on package design and winery size. However, some general statements about cost can be made. The cost of a bottle will vary from $0.30 to $2.30. The label printing costs can be from $0.10 to $0.85 each depending on size, content, and quantity. Engraved labels are economically feasible only in runs of less than 1,000 or a million or more.

Design costs vary upward from a minimum of $20,000, unless the winery can find a way to work directly with the printer. In this case, either the winery or the printer will have to design the label. Larger wineries, using complex design processes, can spend $150,000 to $200,000. A design firm will often begin proposals with five phases of design, the last of which is completion. Each phase is devoted to a refinement of the beginning idea and treats such things as whether the art form is hard copy or digital. Some labels have required as many as 25 phases, each lasting two or three weeks. Most work done by my firm is in the $25,000 to $65,000 range. These types of projects are based on in-house or tasting room customer research. If conventional consumer research is required, this will add substantially to the cost.

Credible market research is a good investment for large brands with significant revenue flows gained by each percentage-point increase in sales. Two types of research are recommended. The first identifies the equities in the existing label, and the second tests the proposed new label to determine its relative effectiveness. If the new and old labels have close to the same impact, market experience indicates that it is preferable to adopt the new one.

There are two research approaches that can be effective. The first is to make a model or dummy of the product and show it to consumers in the tasting room or another controlled environment in which questions can be asked. There may be more than one model, and it is not necessary to indicate the correct brand name. The objective is to learn what conclusions about a product's attributes a consumer will draw after seeing the model. These attributes are compared with the reality of the product to make sure that the chosen model clearly conveys the truth. An alternative approach is to ask consumers what they like and dislike about the old and new labels. The label that makes more consumers want to buy the product would then be chosen.

The research methodology may involve one-on-one interviews, focus groups, or other retail intercepts. The intercept process involves watching what people buy and then asking them about all the factors that influenced their purchase decision. This type of research is complex. The research design may lead to the right questions and the correct interpretation of responses. But it can also lead to biased results when the questions are not carefully asked or the responses not properly classified. A common problem is the bias introduced by asking leading questions, that is, questions that are phrased in such a way as to signal a preferred answer.

CONCLUSIONS

There are several important lessons to be learned about label and bottle design. Perhaps the most important is that design changes do help maintain or enhance a winery's competitive position. In wine stores, one can observe the multitude of new designs and how they change over time. It is clear that wineries, facing an increasingly competitive environment, have reached out to designers to help the wineries gain a competitive edge. Design changes would not occur if they did not benefit the wineries. The second lesson is that changes in design require a significant amount of thought and planning. A design should not be based on a drawing on the back of an envelope and inspired by a passing thought. Label and bottle design should be compatible with other elements of the winery's advertising and promotion scheme. Finally, change costs money. Any decision to differentiate the winery's product through changes in labels and bottles should be based on reasoned economic analysis. In other words, they should promise to be profitable. With these lessons in mind, a marketing manager should be prepared to use label and bottle design as part of an ongoing marketing strategy.

CHAPTER 18

Using the Bulk Wine Market as a Marketing Strategy

Steve Fredricks

INTRODUCTION

To many people, wine marketing is the process of selling bottled wines through the distribution chain to the ultimate consumer. This process incorporates strategies of product differentiation, production and inventory management, pricing, distribution, and promotion. The concept is too narrow because it ignores the strategic use of the bulk wine market. For example, a winery can gear its entire marketing strategy to selling on the bulk market, eliminating the need for bottling, distribution, and related activities. Or a winery can source its product entirely on the bulk market, eliminating the need for production facilities and allowing it to concentrate on marketing. Or a winery may use the bulk market to stabilize production and pricing by purchasing bulk wines in short years and selling bulk wines in surplus years.

It is surprising to learn of the volume and types of wines that are sold in bulk from one winery to another, and how many wineries throughout the world use bulk wines in their own blends. Some wineries do not produce a drop of the wine sold under their own labels and may not even bottle it themselves.

This chapter analyzes how the bulk wine market can be used as part of a marketing strategy. It considers the services offered by bulk wine brokers,

Steve Fredricks received his BS in wine production management and marketing from UC Davis. He is vice president and partner in Turrentine Wine Brokerage, one of California's major bulk wine brokers.

the process of making transactions within the bulk market, the forces affecting the bulk market, the ways in which wineries use the market, and the importance of the wine market cycle. Its purpose is to broaden the concept of wine marketing.

BROKER SERVICE

A bulk broker provides various services. The principal one is matching buyers and sellers, which range from multimillion-case wineries to 200-case private-label blends for restaurants. Brokers provide advice on market trends and pricing, credit information on prospective buyers and sellers, and information on the hundreds of other deals going on in the market. They provide convenience through instant accessibility, full services, and attention to the details and documentation of the sale. Hopefully, the least needed service is the mediation of disputes. The experience of my firm is that 99 times out of 100 everyone is happy with the deal, but there is always that one time when someone says that the wine does not taste exactly like the samples and this requires mediation.

The actual transactions of buying and selling bulk wine are rather informal. Sellers list a wine for sale with a broker, and the broker advises on current market prices and trends and an asking price. Prices, of course, can vary. We had a client in the winter for whom we were able to sell chardonnay at $20 per gallon, but when the client returned in June, the best price we could get was $14 per gallon. Once the winery decides to go ahead with the broker, it will send samples. The broker will review its database and send samples out to most current buyers for those wines and to those who bought that wine from the same producer in the past.

If a buyer likes a wine, he or she will call and make an offer and the broker will negotiate an agreement. In addition to the price per gallon, the negotiation includes payment and shipment terms. The standard terms are for shipment within 30 days of the date of the agreement and payment within 30 days of shipment. There are times when buyers may be required to put down a deposit or pay the total amount due in advance of shipment. Most bulk wine agreements are not signed contracts, just verbal commitments between buyers and sellers, and because of the volume of wine shipped under these verbal commitments, people treat them very seriously and they are very rarely broken. Typically a seller pays the broker a 2 percent commission on sales when the buyer pays the seller.

HOW THE BULK MARKET WORKS

The bulk wine market is driven by supply and demand, although consumer demand appears to be the most powerful driver. Consumer preferences for chardonnay, white zinfandel, and merlot drove bulk prices for these wines considerably beyond the equivalent prices for grapes paid by wineries. For example, in marketing year 1996/1997, bulk cabernet sauvignon from Napa and Sonoma Counties was selling at $25 a gallon, the equivalent of $4,000 per ton, but the spot market prices for cabernet sauvignon grapes were mostly $1,800 to $2,200 per ton during that same time.

A short harvest also pushes prices up. In 1994 central coast chardonnay sold for $10 to $12 a gallon. In 1995, a year of short harvest, prices went up to $19 a gallon, only to fall to $13 to $15 in the following year after a better crop. These variations were more pronounced than variations in wine grape prices. The quality of the crop also affects bulk prices more quickly than grape prices. The bulk market also fluctuates with the economy. The economic boom of the 1990s corresponded with a steady climb in average bulk wine prices. This reflected the higher prices that wines were able to command on the consumer market relative to the preceding decade.

The global market for wine has not had much of an impact on California bulk wines, although it is likely to in the future as California production expands and imports struggle for market share. In the mid-1990s, California bulk cabernet sauvignon, merlot, and chardonnay were in short supply and very expensive, even though the same varietal wines were available in bulk from France and Chile at a better quality and lower price. When the inevitable oversupply of California wine and grapes arrives, the global market for wine will have a greater influence. California growers will likely face lower prices and will need to improve quality to be on par with wines from Chile and the south of France.

THE MARKETING ADVANTAGE IN USING
THE BULK MARKET

Most every winery uses the bulk market to some extent. Some large Central Valley wineries such as Delicato use the bulk market as their main business; it is the principal focus of their marketing program. These wineries are able to avoid major marketing investments and brand development costs. Ninety percent of the wine produced by some of these wineries is produced under a long-term contract to other California wineries, to wineries in other states, or to wineries elsewhere in the world. They have become experts in efficiently producing wine to meet buyer specifications. These wineries account for the majority of wine sold in bulk.

Although a few large coastal wineries produce predominately for the bulk market, more of them are buyers of bulk wines, using them for blending and augmenting their own production. This practice keeps them in touch with bulk wine prices and allows them to exploit opportunities to keep their own production costs down. Some wineries also may produce more than their needs in order to pick the best wines for blend and then sell off the excess in the bulk market. This also allows them to use all of their production capacity to spread out their fixed costs. From a marketing standpoint, these wineries are gaining stability in product availability and quality control.

The small and midsize premium wineries gain market advantages by using the bulk market mainly to fine-tune blends with bulk wines from other regions with different characteristics. They use it to sell off lower-quality wines or to purchase lower-priced wines for second or private labels. They also use the bulk market to increase production beyond their processing capabilities or their vineyard capabilities. These uses help maintain a consistent supply of wine with dependable quality characteristics.

UNDERSTANDING WINE MARKET CYCLES

Marketing decisions vary according to the phase of the wine market cycle. The phases can be characterized as follows:

- phase 1: rapid sales growth and increasing margins
- phase 2: good sales growth but narrowing margins
- phase 3: drop in growth rate and disappearance of margins
- phase 4: slow growth rate and reappearance of narrow margins

These phases do not necessarily follow one another in order; this is particularly apparent when growth rates and prices are affected by large swings in grape and wine production.

Phase 1 is characterized by rapid sales growth and increasing margins. Strong demand allows wineries to eliminate discounts on case goods and increase prices. Wineries gain increased margins because they are selling products produced from low-priced grapes in phase 4. In phase 2, sales growth is strong but restricted by tight supply and higher prices. Margins are narrowing for the wineries at this time because grape and bulk wine prices have been increasing faster than case good prices. In phase 3 the growth rate drops and margins disappear as supply exceeds demand. A change in the economy may soften consumer demand to exacerbate the situation. Consumers resist prices just as new supplies of wines and wine grapes are flooding the market. Wineries have to discount case goods prices and spend more on promotion in order to move excess inventory, and their cost of goods sold during this time is higher because it was acquired during phase 2, a time of rising grape prices. Winery margins shrink or disappear. Phase 4 features a slow growth rate and the reappearance of margins. The case goods market has become very competitive, so there is a growth in sales, but the growth is very slow relative to the amount of supply out in the market. Narrow margins reappear because supply exceeded demand for grapes and bulk wine, so grapes and bulk wine prices have declined and reduced the wineries' cost of goods sold.

Phase 1 is the best for the winery. The winery is highly profitable because sales and prices are up but costs are low, reflecting purchases during phase 4. Phase 3 is the worst for the winery, which experiences an increase in supply and a weakening in demand. Costs are high because grapes were purchased under long-term contracts signed during phase 2, when there was crazy demand and crazy prices followed. The winery's competitors face the same oversupply situation and react by discounting prices in order to move inventory.

Table 18–1 illustrates how grape and bulk wine prices might differ and what this indicates about supply conditions. Under a balanced supply and demand situation, the price for bulk wine should be about equal to the grape price plus the cost of processing. However, in times of surplus supply, bulk wine prices tend to drop more rapidly than average grape prices, and the reverse happens when supplies get tight. The example is based on bulk wine prices in the spot market and the average grape price paid for the variety in question. Bulk prices have been converted to equivalent per ton prices at the rate of 160 gallons per ton. The exact figures are not as important as is their usefulness in indicating market conditions.

It is apparent in the table that prices for the grape variety increased between 1991 and 1992 while the per ton equivalent of the bulk wine price decreased. This indicates an oversupply. The situation began to correct

Table 18–1 Comparison of Average Grape and Bulk Prices, 1991–1997, in Dollars per Ton

Year	Average Grape Price	Bulk Price	Difference
1991	1650	1440	210
1992	1700	1280	420
1993	1520	1312	208
1994	1400	1328	72
1995	1560	1888	−328
1996	1700	2368	−668
1997	2500	4000	−1500

itself in 1993, when the grape price declined and bulk prices rose a small amount. The difference between the two prices narrowed. Market balance appears to have been achieved in 1994, when both prices were nearly equivalent. However, after that, demand began exceeding supply, and bulk prices increased at a substantially faster rate than grape prices. The difference between the prices became negative; that is, the grape equivalent price of bulk wines exceeded the average grape price. The table indicates that 1991 to 1993 was a period of excess supply and 1995 to 1997 was a period of excess demand.

In terms of the market cycle, the market was moving out of phase 2 and into phase 3 because it moved into a time of excess supply where the bulk wine price was lower than the cost of grapes plus processing. In 1996 it moved back into phase 2, where it stayed through 1997. The situation changed a little for some varieties. For example, bulk merlot that sold for $25 to $30 per gallon in 1997 was selling for $18 to $20 in 1998. This suggested an increase in supply that was likely to depress grape prices.

CONCLUSIONS

There is active trade of bulk wine in California and in the world. The bulk wine market provides marketing opportunities for participants. As this chapter has shown, one of the most prominent advantages of the bulk wine market is in allowing wineries to maintain consistent supplies and blends over time when grape supplies vary. In times of shortage it is possible to use overseas bulk wine supplies, and in times of surplus it is possible to move excess production away from targeted markets. The bulk wine market is also used to maintain second and private labels. Information from the bulk market is useful to all wineries and growers for long-range planning.

The bulk wine market operates quickly and relatively efficiently in matching buyers and sellers. It also provides signals, through relative prices, of market supply situations that may not be as readily apparent through proprietary pricing arrangements in the bottled wine market. For example, comparative prices can be used to estimate in which part of the marketing cycle the market is.

It is important that wine marketers understand wine market cycles because strategies vary according to market phase. Phase 1 is the best for wineries because sales are growing, prices can be advanced, consumers and trade are receptive to promotion strategies, and costs reflect the low grape and bulk wine prices of the last phase. The worst phase is the phase 3, where demand has softened and supplies continue to grow. Prices tumble, grape costs reflect the more buoyant previous phase, and margins are inadequate to permit advertising and promotion. Careful analysis and use of the bulk market can enhance the marketing strategies of wineries in each of these marketing phases.

PART IV

Profiting through Positioning

CHAPTER 19

Basics of Brand Positioning

Richard A. Gooner

INTRODUCTION

The objectives of this chapter are to define what is meant by brand positioning, to determine where it fits in the logic of marketing, and to illustrate how it can be implemented.

BRAND POSITIONING

A common misapprehension is that positioning relates to where a brand is placed on retail shelves; in essence, shelf placement is a product-focused strategy, and it is important. More important, however, is the positioning of a brand in the mind of the consumer. In essence, positioning is a consumer-focused strategy. The first part of positioning involves assessing where the consumer already has positioned the brand relative to other brands. The second part is to change that position through marketing strategies to increase the likelihood of brand purchase. The marketer may seek to shift a brand to fill gaps in the positioning of other brands by the consumer, for example, to fill the need for a "value" brand or a "refreshment" brand.

Positioning is an objective of a marketing strategy. Therefore, all elements of the marketing mix must be consistent with the positioning target. This includes decisions about product characteristics, packaging, price,

---

Richard A. Gooner received his PhD in marketing from the University of North Carolina and currently teaches at the University of Alabama. Prior to returning to academia, Richard was vice president of sales and marketing for Glen Ellen winery and was vice president of international brand development for Brown-Forman.

promotion, and physical distribution. Inconsistency will damage a marketing program by sending the wrong signals. A winery with fancy or elegant packaging for a $4.99 wine is wasting money and sending a misleading quality signal to the consumer. On the other hand, if the wine sells for $11.99, the packaging should reflect the high quality of the wine and the care with which it has been made. About the only expensive wines that can get by with unexceptional packaging are the really well known international wines. In some cases, the product may be the driver and the positioning objective will be to place the existing product in a favorable position. In other cases, the existence of a gap in the market may be the driver and the objective will be to design a product, package, and distribution system that will fill the gap.

The development of a positioning plan depends on knowledge about what position the brand currently "owns," what position the brand ought to own, who the competition is, and what resources are available for positioning. The plan should also take into account the match between the seller's marketing mix and the intended (or actual) brand position. In considering the marketing mix, wineries should recognize that positioning is a strategy for differentiation. It seeks to establish a difference in the consumer's mind so that the brand is clearly different and preferred for satisfying consumer needs and wants. If a seller does not have effective differentiation, then it had better have an unbeatable price.

Differentiation through product positioning and other strategies is often used in conjunction with market segmentation. Segmentation and clustering is the process of reorganizing parts of a market into groups of consumers that share more similarities with members within the group than they do with people outside the group. The objective is to cluster consumers based on their needs or based on their receptivity to a given message. Thus, marketing strategies can be designed to fit the characteristics of the individual clusters more effectively than would a mass-marketing strategy aimed at the total market. The segments have to be different enough that they are receptive to different positioning strategies. Segmentation may also involve regrouping smaller segments into larger homogenous markets. However arrived at, the segments need to be profitable and actionable. That is, marketing efforts to position the brand must be able to reach the segment at a feasible cost. Usually, the two strategies go hand in hand, with a differentiation strategy, including product positioning, crafted to meet the characteristics of the newly defined market segments.

Consumers are continually reappraising brands as new competitive conditions emerge. This continual reappraisal often creates the need, and opportunity, for new products and brand repositioning. But it should not obscure the fact that even the most aggressive new product development firms still get most of their sales from "old" products. The challenge is to actively and explicitly manage brand repositioning and new product development. The three following examples illustrate this process.

I was the brand manager for André champagne some years ago when it was selling for between $1.99 and $2.29 per bottle at retail. A question was raised about repositioning the brand against some of the newly available brands, such as Freixenet, that sold at higher prices. In my opinion this would not work, regardless of the investment, because the jump in position was too great to be consistent with consumer perceptions. However, it was possible to reposition the brand slightly and with considerable profit. The strategy was to change the André label and move from selling pink champagne, which was a legacy of the 1950s and 1960s, to selling a "blush" sparkling wine at a time when blush wines were becoming very chic. It involved developing the concept of André Blush; changing the label, which cost a portion of a penny per bottle; and raising the price to a consistent $2.29 per bottle. It was a small repositioning, just a tweak, and not particularly brilliant, but the money just rolled in.

The second example relates to M.G. Vallejo. When I was with the Benzinger organization, there was a discussion about the possible sale of the M.G. Vallejo brand for $1 million. This was a brand that had relatively little consumer recognition and sold about 400,000 cases annually in a retail price range of $4.99 to $5.99 per 750-milliliter bottle, with strong sales in the 1.5-liter size. The decision was made to attempt to reposition rather than sell the brand. With this in mind, the organization studied the market and found that chardonnay was the "hot" category, selling for $6.99 to $7.99, about $2 above the M.G. Vallejo level. Retailers liked the profitability of this price point. Competition at this point was not severe and because M.G. Vallejo was not well known, it was not rooted in a given position in the consumer's mind. So the organization changed the label, developed a repositioning plan, and took the plan to the distributors for discussion and possible implementation. The concept was simple: a very chic black label, a good product, a new and higher pricing point, and full cooperation with distributors. The company was very successful in enhancing the value of the M.G. Vallejo brand, which was ultimately sold to other buyers for several million dollars. This example demonstrates that repositioning does not have to change everything about the product and its packaging to be successful. It also demonstrates the value of working with distributors at an early stage of the process.

A third example illustrates the successful repositioning of an old product into a new popular class of blush wines. A major producer faced competitors that were producing and selling white zinfandel and created the rush toward blush wines. The producer had large quantities of grenache grapes available that had been used to produce wines that were of declining popularity. It did not have access to sufficient quantities of zinfandel to support a large marketing program. An important consideration was that the price for grenache grapes was one-third to one-half that of zinfandel grapes. The producer named the brand White Grenache and produced it as a pleasant-tasting blush wine. Its cost was far below that of competing white zinfandels. The repositioning resulted in increased profits for the producers and others in the distribution chain. The basis for this success was the observation that consumers, especially white zinfandel drinkers, were sufficiently unsophisticated that they would accept a product called White Grenache that had the same blush appearance and a similar, but not identical, taste. This is an instance where production and finance considerations led to successful product positioning.

THE POSITIONING PROCESS

Positioning is iterative, a constant process of anticipating changes in the product, competition, and consumer preferences. A seller cannot assume a positioning effort is completed just because it has reached the original targets. The seller must follow up to anticipate competitor and consumer reactions and also verify that distributors are continuing to train their people to communicate the positioning message.

It is sometimes useful to think in terms of high-tech and low-tech approaches to brand positioning. Processes appropriate to these approaches are outlined below. But underlying either one has to be an understanding of how the production/marketing system works. Analysts think of this in terms of a model, but managers are more likely to think in terms of understanding. Regardless of the approach to positioning brands and creating a marketing program, there is no substitute for thinking about the consumers, the linked system that serves them, and both present and likely future competitors. Brand positioning strategies need to fill the needs of both consumers and members of the distribution chain. This assumes that the seller knows what these needs are and then can effectively communicate how the positioning strategy will satisfy those needs. Research is an aid in this thinking process, and it may help improve decisions, but only if the underlying research design reflects a thorough understanding of consumer behavior, market organization, and competitive responses.

The low-tech approach relies on talking with tradespeople and customers to develop ideas about how the product might be positioned. Such discussions help wineries think through the alternative positioning strategies. Interviews should be conducted with both people intimately involved with the brand and people who are less familiar with it. This provides some assurance of being in contact with potential consumers as well as current consumers. These results should be supplemented with available demographic and consumer behavior data (income and wine consumption data, for example) to provide a more accurate picture of market segments. This low-tech approach relies on management experience and intuition, which often bring to light certain relationships and nuances missed by technical analysts. Even the most experienced and intuitive managers must think the matter through carefully, however.

The low-tech approach may also involve focus group interviews and perceptual mapping. Focus groups are gatherings of consumers to talk about certain marketing and consumption practices under the guidance of a moderator. Such groups can be incredibly valuable, but assembling them is quite expensive. To get the most out of the experience, the organizer or sponsor of the panel needs to have an idea of what key consumer needs are and what factors influence their behavior. This information will help the moderator guide the discussion and help the sponsor interpret results. The sponsor may observe and listen to the discussion from behind a one-way mirror. This reduces the chances that the sponsor will influence results.

Perceptual mapping is defining where consumers in a particular market segment place a given brand among given criteria. For example, a soap manufacturer talked to people and learned that two key factors influencing preferences for bar soap were whether the soap was low moisturizing or high moisturizing and whether the soap was deodorant or nondeodorant. Important factors like these can be determined in a focus group by asking "What are the factors that really make a difference?" Once the criteria have been identified, the moderator can ask the focus group where the product in question fits in reference to the criteria. This question will show where the products fit on the consumer's perceptual map. It allows the seller to see places on the perceptual map where its product might be positioned. This, in turn, might suggest the marketing mix elements needed to accomplish the positioning. This procedure is low tech and accurate, but it requires a lot of careful thinking and a degree of sophistication to make it work. For example, a manager may disagree with a consumer's assessment of a particular product, and, therefore, where it is placed on the perceptual map. But most success comes in working from what people believe to be true.

The high-tech approach to positioning uses technical tools such as surveys, factor and conjoint analysis, cluster analysis, and perceptual mapping using statistical programming. The approach starts with thinking: subtle and intuitive knowledge of the consumer—not just what the wine maker feels or what the manager feels, but what can be learned by being really in touch with the consumer. This knowledge may come from focus groups, distributors, or retailers, and from qualitative analysis that helps identify what key factors influence consumer perceptions and buying behavior. The high-tech approach will often involve various survey instruments to identify what factors are most important to buyers. Price, perceived quality, and packaging are examples of key factors. The high-tech approach may also involve a test marketing program to try out various positioning strategies and gauge the results.

The factors are then evaluated using computer software that is extremely powerful and cheap and that can evaluate the weight (or importance) of the key factors in decisions. Factor analysis is one such technique. Another analytical technique is conjoint analysis, sometimes called "tradeoff analysis." It provides a way of looking at various tradeoffs between product characteristics. For example, a 1.5-liter package, a new label, and an eye-level shelf position may provide the best combination of product characteristics from a profit standpoint. But if the choice is constrained to only two characteristics, which combination is most valuable? Conjoint analysis will help suggest an answer to such a problem. Cluster analysis, which follows naturally from factor analysis, identifies clusters of customers with similar behavioral characteristics. As in the low-tech approach, the next step is perceptual mapping. But in this case, more sophisticated and costly techniques are used, including multidimensional scaling, a reliable mathematical tool that can be operated on a desktop computer.

Regardless of the sophistication of the original approach, follow-up is essential. The wine marketer is constantly chasing a moving target and hopefully closing in on it. So there is a continuing need to follow up with target customers and trade partners through interviews and surveys. The follow-up may indicate a need for test marketing to fine-tune alternative strategies.

MAKING POSITIONING WORK

Positioning is as much an art as it is a skill. It requires intuition, creativity, flexibility, and a willingness to work out the logic of a program. There are some considerations and cautions that help people make better decisions. One major manufacturer asked the following questions concerning its positioning strategy: (1) What current consumer decisions about the product will be changed as a result of the firm's positioning efforts? (2) Why will the consumer believe the firm's message? (3) What character does the brand bring to the consumer's mind? For example, if the brand were a person would it be Madonna, or Bill Gates, or Queen Elizabeth? (4) What are the things that should or should not be associated with the product (e.g., wine goes better with natural products than with racing cars)?

There are certain simple things to avoid. One is impatience. It takes time to plan a project properly, and there is a strong temptation to cut planning short and jump right into execution. This is a mistake that has doomed many marketing programs. Complexity should be avoided. Ideas should be kept simple, particularly in communicating to consumers. The problem is that complexity is expected from highly paid executives. Simple seems too easy and not worth the money. "Coke is it" is a simple message that has made the company lots of money. Campaigns that are too cute may not work. For years, Volkswagen had a successful campaign based on the idea "think small," but the California raisin industry's "Dancing Raisins" campaign won awards for cleverness but did not greatly increase sales. Somehow the message did not say "buy me." And a winery should not let the numbers overcome sound marketing. For example, a wine may make a large margin when positioned at $7.99, but other factors may not support such a position. Consumers do not care about producer financial objectives and will not support them unless the entire positioning package is right.

Ad agencies and consultants can be helpful, but they are not a cure-all for marketing problems. Producers should recognize that no creative approach, no matter how clever, is going to turn a wine into something it is not. That is, the product development work must have been completed and the product must actually be what it is claimed to be. A further caution is to make certain that there is enough of the product available for sale after the promotion increases the sales volume.

Producers should keep an outside rather than an inside perspective on product positioning. Having an inside (the firm) perspective may encourage producers to generate detailed, complicated messages describing management-defined product attributes and internal objectives while ignoring consumer realities. To avoid this pitfall, producers should (1) keep the positioning message simple, because simple messages are the most powerful; and (2) maintain a consumer focus. Producers must ascertain what the consumer thinks of the brand and its current position, and what the reaction to a new position is likely to be.

CONCLUSIONS

Product positioning is a management strategy, pure and simple. It requires an understanding of the market, an ability to work with other participants in the marketing chain, a willingness to listen to new ideas, and a commitment to planning, implementing, adjusting, and planning again.

CHAPTER 20

Positioning Multiple Wine Brands

Stephanie Grubbs

INTRODUCTION

This chapter is a case study of multiple brand positioning by a major California winery, Robert Mondavi. Its purpose is to demonstrate how the principles of positioning are used at the practical level. To accomplish this, the following discussion centers on the positioning or repositioning strategies and tactics for three brands, Robert Mondavi Napa Valley, Robert Mondavi Coastal, and Woodbridge by Robert Mondavi. Other brands within Robert Mondavi's portfolio are used to illustrate special points, and comparisons are made with the strategy of another major winery. The chapter is based on my experience with Robert Mondavi and reflects my own views as to the lessons that can be drawn from that experience.

BACKGROUND AND CONCEPTS

When I was in sales for Robert Mondavi in the late 1980s, the company had five brands. I can remember being very nervous when we decided to introduce a sixth brand, Vichon. Since then, however, we have expanded to 12 brands in just under 10 years. This has created a phenomenal challenge in trying to position the various brands and keep them separate in the minds of consumers and the trade.

Stephanie Grubbs, a Napa Valley native, holds a degree in business administration from California State University, Sacramento, and a wine marketing certificate from the University of San Francisco. After a 14-year career in marketing at Robert Mondavi Winery, she is now consulting for several wineries under her own Moss Lane Wine Marketing.

The basic responsibility of any marketing manager is to create demand at the consumer and trade level for each brand. That entails treating each brand as a different product with a distinct personality. We spend a significant amount of time managing or juggling the basic marketing mix for each of the brands. The mix, as well documented in marketing literature, includes price, product, place, package, promotion, and people. Price is a very big issue for our high-end wines like Opus One, but promotion is not as important for Opus One as it is for Woodbridge or Coastal. Place is obviously very different for these wines. Opus One and the Frescabaldi joint-venture wine might be found in high-end wine shops and not in grocery stores. Everyday wines like Woodbridge and more casual social wines like Coastal are definitely grocery-store driven. Packaging (bottle size, label appearance, and other features) varies, to reinforce the consumer image of the wine. It is well known that promotion must be tailored to individual brands and markets. The promotions for Napa Valley will stress different aspects of the wine and the wine drinking experience than will those for the more casual Coastal brand. Finally, the people part of the marketing mix will vary. For example, the involvement of the Robert Mondavi family, or the wine makers, or viticulturists will vary importantly according to the brand and its market requirements, or the need may be for brand generalists rather than specialists.

Those of us at Robert Mondavi have been spending a lot more time talking to the consumer and not just focusing on the trade, as in the past. We have learned that in order to position our expanded number of brands, we need to know more about consumer preferences and behavior and not rely solely on what the trade thinks. We have made a dramatic shift by investing relatively more money and resources in the pull part of the marketing mix rather than the push.

Part of the reason that we are emphasizing consumer communications is that the message is becoming complicated. Because many of these complicated messages go through the sales force to the distributor, and then to the retailer, and finally to the consumer, there are ample opportunities for the message to be garbled and misunderstood. This is true also in dealing with the media, where the message may go through our marketing department, then through an agency, on to a writer and editor, and finally to the public. The market is crowded with competitors, making it essential that the message get through clearly and unambiguously. Robert Mondavi finds it best to deliver a simple, unadorned message to the consumer. To be effective, regardless of the delivery method, the message must be clear both to the consumer and to the people creating the message.

BRANDS AND POSITIONING

Branding philosophy plays an important role in managing multiple brands. For Robert Mondavi, this philosophy changed from a belief in using the Robert Mondavi name on all brands to a belief in using the name on relatively few. The previous strategy extended the Robert Mondavi name to new brands; in effect, the winery was introducing new line extensions. Coastal and Woodbridge were both line extensions of Robert Mondavi. Consumer research in the early 1990s determined that consumers were unclear as to how these wines should be positioned in their own minds. As a result, the winery changed the Woodbridge wines, which were formerly labeled as Robert Mondavi table wines (red or white), into a stand-alone brand called, simply, Woodbridge.

The Coastal brand was introduced to capitalize on consumer demand for moderately priced premium wines, with grapes sourced from coastal growing regions. It would allow the winery to reposition the Napa Valley brand toward higher-priced prestige wines of limited availability. It would also alleviate the pressure to have Napa Valley or Opus One brands in the club stores and the grocery chains. The positioning strategy was to create images in the consumer's mind of separate but complementary brands. However, when the Coastal brand was introduced, its name and label were close enough to those of the Napa Valley wine to create confusion among consumers. They were unsure of which Robert Mondavi wine they had. An alternate strategy of introducing the Coastal brand under the name "Brand X by Robert Mondavi" might have been better. This would have reduced confusion and eased the task of positioning the brand. There is a tradeoff, of course. The use of the Robert Mondavi name allowed a rapid buildup in sales at a lower cost than would have been required for a new brand, even with the Robert Mondavi endorsement. However, this saving might have been offset by the costs of consumer confusion, which could dampen future growth, cause losses in sales of the Napa Valley brand, and lead to defection to other brands.

The winery discovered that individual stand-alone brands, like Opus One and Woodbridge, with which Robert Mondavi's name was not closely associated, had much clearer positions in consumers' minds. It was decided that new brands would have the Robert Mondavi endorsement but would not rely on the Robert Mondavi name in the name of the brand. For example, a new brand produced in a joint venture with the Chadwick family in Chile has the endorsement of both the Robert Mondavi family and the Chadwick family but is called Caliterra. The objective is to build a completely separate position in the minds of the consumer by using a separate brand name.

A comparison between three Robert Mondavi and three Wine World brands illustrates how the old strategy differs from a stand-alone strategy. The Robert Mondavi brands were Robert Mondavi Napa Valley, Robert Mondavi Coastal, and Woodbridge by Robert Mondavi. The winery had very clear, distinct positions for these brands in that Woodbridge was the everyday wine, the Coastal brand was more of a social or casual wine, and the Napa Valley brand was a higher-end wine for special occasions. Wine World's answer to each of these brands was a completely distinct brand name with its own personality. The company achieved clear positions for Beringer, Napa Ridge, and Meridian in the minds of consumers. Consumers knew what to expect, where the brands fit, what price points were appropriate, and when to use the brands. Beringer was positioned against the Napa Valley brand, Meridian against the Coastal brand, and Napa Ridge against Woodbridge. There was no obvious link to Wine World, or between the original Beringer brand and Meridian or Napa Ridge. They built these as completely separate brands. The new Robert Mondavi brand strategy is closer to the Wine World model than to the old Robert Mondavi strategy.

The difference between related brands and stand-alone brands is evident in other product areas as well. Nike and Mercedes-Benz follow a strategy of brand extensions, and Procter & Gamble utilizes independent brands. Nike and Mercedes-Benz use their names on everything they make. They may have a model number that is different, or a name of a shoe that is different, but the company name and logo are clear on all the brands that they are trying to sell. This strategy appears to be successful for them; the names Nike and Mercedes-Benz connote high quality. This was the strategy followed by Robert Mondavi until consumer confusion about different product categories dictated a switch.

Robert Mondavi switched more in the direction of Procter & Gamble, a company that builds separate brands. One has to study carefully a toothpaste, soap, or shampoo package to find that the product is made by Procter & Gamble. This strategy has been extremely successful for Procter & Gamble.

APPELLATIONS

The use of appellations presents some interesting marketing challenges for Robert Mondavi. The winery's guiding philosophy has been to produce the finest wines from the finest regions, searching for the perfect match between variety, soil, and climate; all this boils down to appellation. For a large percentage of our consumers, especially at the Woodbridge and Coastal levels, appellation does not mean very much. These customers are not really concerned about where the wines come from. So an appellation philosophy is based on something that does not mean much to probably two-thirds or three-quarters of Robert Mondavi's customers.

The task for the winery is to convey to consumers what appellation means in terms of wine characteristics or attributes so that they can use this information in positioning our wines. If we are successful, appellation will become meaningful to consumers of at least the Napa Valley and Coastal brands. For example, we believe that the central coast gives pinot noir and chardonnay the best expression of fruit in the wine. This seems irrelevant to a consumer who does not want to know much more than the variety and the price. For us, as producers, the incredible fruit expression from the central coast is the reason that we have the wine in our portfolio. Trying to make that mean something to the consumer is a challenge.

The Robert Mondavi branding and positioning scheme is based on appellation. Woodbridge wines, which have a California appellation, can be sourced from anywhere in California, although we have primarily selected the Lodi region. The positioning statement is that these are barrel-aged wines for every day. The wines are priced at about $6 to $8 per 750-milliliter bottle (Table 20–1).

The Reserve wines have the most narrow appellation, one that encompasses designated vineyards in Napa Valley. These wines are the most expensive and come from the narrowest sourcing that we have. The District wines are from recognized Napa Valley subappellations, such as Carneros. The volume is higher than that of Reserve wines, and the price is lower. The next level is the well-known Napa Valley brand designation. Volumes are greater than in the preceding two classifications, and prices are lower. Robert Mondavi is seeking to reposition its Napa Valley desig-

Table 20–1 Mondavi Wine Brands, Sources, and Prices

Brand Designation	Source of Grapes	Estimated Retail Price
Mondavi Reserve	Napa vineyard designated	$25–$55
Mondavi District	Napa subappellations	$22–$25
Mondavi Napa Valley	Napa Valley	$18–$22
Mondavi Coastal	North and central coast	$8–$14
Woodbridge	California	$6–$8

nation at a higher level. The Coastal designation has higher volumes and lower prices than the Napa Valley designation, and its sourcing is far less restricted.

SALES AND TRADE STRATEGIES

The key strategic decisions in marketing and positioning multiple brands concern how to organize the sales force, what distribution network to use, which brands to present to which parts of the trade, and how to influence the consumer. Creating an effective internal sales force is a major task. Should the sales force be composed of brand generalists, each with the responsibility for representing all Robert Mondavi brands to their clients? Or should the force be composed of brand specialists calling on the same clients as other specialists? As the number of brands in the Robert Mondavi line increased, the sales force was put under tremendous pressure to compete for internal marketing resources. Robert Mondavi decided to have a single internal sales force representing all of its brands before appropriate clientele but to expand the force by including specialists in education and training, and segment specialists in chain stores.

For the most part, all of the Robert Mondavi brands are handled by the same distributor. The exceptions tend to be with our joint ventures, Frescabaldi and Opus One. This is important from Robert Mondavi's perspective because it allows us to develop a set of brand priorities that can be followed through the distribution process. Distributors gain from dealing with a single "expert" firm concerning a set of related brands. In any case, distributors face the same problems as does Robert Mondavi in continually evaluating the sales priority for each brand and keeping the sales force focused on brand and image building rather than just short-term case movement.

When a winery acquires existing brands, it must determine whether to link them back to the "parent" brand. The most influential factor in this decision is generally the independent strength of the acquired brand. Strong brands can more easily stand alone, at least in consumer marketing. Robert Mondavi differentiated between the preferences of consumers and those of the trade in adopting an identification strategy for its acquired brands, Vichon and Byron. Recognizing that these brands had a strong consumer franchise, it chose not to identify a link to Robert Mondavi. The winery concluded that such a link would not add anything to an already well-recognized brand personality. But at the trade level, the attraction of the Robert Mondavi name for service, value, and stability made it important to include the acquired brands as part of the Robert Mondavi family of wines. So at the trade level, Robert Mondavi markets all of its brands together, but at the consumer level, it focuses on the individual brands.

CONSUMER COMMUNICATIONS

Each of the brands is identified separately in consumer communications. Woodbridge, Coastal, and Napa Valley have individual advertising and promotional budgets aimed at developing each brand's own personality and avoiding confusion with the other brands. The success of these programs depends on market research that identifies consumer characteristics and how buyers of one brand differ from buyers of another. An important research objective is to learn how the winery can appeal to the consumers separately in all of its campaigns. The result is a series of consumer profiles for each of the brands that identifies consumer demographic characteristics, where they shop, how and when they use wines, how loyal they are to wine, what wine types and wine brands they like, how they learn about wines, how much they want to know about them, and what influences their purchase decisions. These profiles differ for various brands and price points, and there is substantial cross-over between brands, caused by the fact that there are different occasions for wine use. For example, the average consumer may drink Woodbridge with pork chops and rice during a weeknight dinner at home, welcome in the weekend by purchasing a couple of bottles of the Coastal brand for a casual dinner party, and select the Napa Valley brand when entertaining the boss.

This one consumer is included in the profiles for each of the brands, with the purchase decision dictated by the occasion of use.

In all of our communications, from brand to brand, the focus is on the consumer, but the winery has to continue educating the trade about what makes Robert Mondavi different. Although the consumer of lower-priced brands is less interested in appellation than in brand and price, Robert Mondavi is building its reputation and its reason for existence on its appellation story. Consequently, it is very important that the trade and the press understand that story. So while a lot of the marketing efforts are aimed at the consumer, Robert Mondavi is not taking its eye off of the trade just yet.

BRAND LOYALTY

There is a pretty clear line between the amount that one pays for wine and the level of loyalty to the wine. When we were doing some consumer research for Coastal radio advertising, we found that the more a person pays for wine, the less loyal he or she is to the brand. People who buy more expensive wine select from a huge range of brands. They actually like the adventure and like to be cutting-edge. But the less people pay and the more everyday their brand is, the more loyal they are. There were some women in our survey who were adamant about having their glass of Woodbridge every night; they viewed it as a reward and were extremely loyal to the brand. But at the higher Coastal price level, there are five or six brands that consumers will choose from, depending on which one is on sale or which one is displayed. The choices are even greater at the prices of the Napa Valley brand. For example, if a consumer went to the store to buy Robert Mondavi Reserve or Opus One and the wine was not on the shelf, the buyer would easily shift to another brand. Robert Mondavi likes to think that it has a higher level of loyalty because of the heritage and longevity of the name, but at these prices, people are much more adventurous and much more willing to experiment. This is the reason that we see significant efforts by high-end wineries to further distinguish their product by using limited appellations (e.g., districts or specific vineyards) and restrictive practices such as estate bottled or proprietor's reserve. The challenge for a multibrand winery is to develop marketing programs that capitalize on these differences in brand loyalty.

INTERVIEWS

Determining consumer preferences through interviews is risky at times. We have found that consumers tend to claim they drink a lot better wine than they actually do. We did a consumer survey across all the Robert Mondavi brands asking if the respondent had drunk a Robert Mondavi chardonnay in the last three months. Of the people who responded positively, 38 percent thought that they had drunk Robert Mondavi Chardonnay Reserve, even though that wine is only 5 percent of our production. If the response were taken at face value, it would mean that consumption exceeded production by a large margin. Because of this sort of bias in survey results, and because of the cost of surveying large populations, we tend to use surveys carefully, and in conjunction with expert opinions and experiences of our own staff and those of the trade. For example, for the Coastal advertising campaign, we used the results from 10 focus groups of roughly 10 persons each. This is a rather small sample size, and we felt it necessary to include additional factors, such as the results of consumer behavior studies or the opinions of selected retailers, in establishing our advertising program.

MARKET PLANNING

The following six steps are essential to market planning:

1. writing mission statements for the company and the brand
2. composing positioning statements for the brand
3. analyzing consumer profiles for the brand
4. establishing brand objectives
5. devising strategies for achieving brand objectives
6. designing tactics for implementing strategies

The first step is to review the company mission statement and then articulate brand mission statements that are compatible with the company statement. The next step is to analyze the consumer profiles that have been compiled to determine what sorts of objectives and market strategies might be appropriate. Once the general objectives and strategies have been developed for each brand, then specific tactics for a marketing campaign can be detailed.

The following illustrates, at least in outline form, how this stepwise procedure was applied to the Napa Valley and Woodbridge brands at Robert Mondavi. The Robert Mondavi company mission is to be the world's preeminent wine producer. The mission for the Napa Valley brand is to be internationally recognized as the world-class quality and image leader in fine wines. So it fits within and supports the company mission statement. The positioning derived from the mission statement is that Napa Valley brand wines are premium wines hand-crafted from the best fruit available in the Napa Valley. There are three tiers within this brand, Napa Valley, District, and Reserve wines, and they highlight different styles of wine making and the personality of Napa Valley's distinctive microclimates. The brand integrates wine into an overall lifestyle that includes enjoyment of food, the arts, and cultural enrichment.

Global sourcing is consistent with the mission of worldwide preeminence and has become part of Robert Mondavi's marketing strategy. Its objective is to obtain the best match between climate and variety for producing wines that fit within the Robert Mondavi array of brands. The winery has entered joint ventures with the Frescabaldi interests in Italy, the Chadwick family in Chile, and other producers under the Vichon Mediterranean label. The mission and positioning statements are developed as described for the other Robert Mondavi brands. They must be consistent and compatible with those brands in order to communicate clear messages to the trade and consumers.

Woodbridge is a moderately priced quality wine, classed among the "fighting varietal wines" that make up the largest part of premium wine sales. Its mission is to be the recognized quality leader in wines for everyday enjoyment. Woodbridge wines are positioned as barrel-aged wines that wine lovers everywhere can enjoy daily. To achieve this position, obviously, the brand must be marketed in supermarkets, clubs, and other chains that offer widespread distribution. Robert Mondavi believes that all its brands should be at the high end of their respective categories in quality and price. Consequently, Woodbridge is encouraging people to trade up from within its category as well as from the lower-priced jug wine category.

EXPANDING DEMAND

The strategy of persuading consumers to switch from other brands to Woodbridge begs the question of how "new" wine consumers can be attracted. Traditionally, no individual member of the wine industry has taken on the challenge of expanding wine consumption because it would be so expensive. It is much more expensive to try to steal away a micro-brew drinker or an Absolut drinker or a Coca-Cola drinker than it is to shift somebody between Kendall-Jackson, Robert Mondavi, and Gallo. The problem is how to best invest advertising and promotion budgets that are relatively small compared with those used to advertise and promote drinks in other categories. It is probable that Gallo and other producers with very extensive wine advertising programs can singlehandedly attract new consumption for all wines in a category even though their purpose is to build only their own brands.

Other organizations, both private and public, have sought to expand demand for wines from specific places, such as California, Bordeaux, Rioja (Chile), or Germany; or to expand demand among younger adults. The first type of positioning activities has met with varying degrees of success in increasing sales for the targeted wines, but their success in adding to total demand, rather than just changing market shares, is less certain. The second type of activity is based on educational programs that focus on wine's place in a healthy lifestyle. It is clear that some of these efforts have paid off in either expanding demand (red wine, for example, after the *60 Minutes* broadcasts about its health benefits) or decreasing the rate of decline. There is continuing debate about the "best" way to organize generic campaigns. In some countries, the preference is for some sort of public agency, such as a marketing board, commission, or professional group. Such agencies have the equivalent of taxing power, so that all producers covered by the agency must pay to support it, either on their own or in combination with public funds. Producers in other countries prefer voluntary organizations, such as trade associations or councils, that carry out programs that are paid for by cooperating members. The "free-rider" problem emerges under voluntary arrangements because nonpaying producers may benefit as much as paying producers. However the efforts are organized, individual marketing managers must still determine how to position their own products successfully within the environment created by the generic wine promotion efforts.

BRAND PLANS

Let me return to the Robert Mondavi example and consider how we repositioned our product lines in a new environment. In this case, it was an environment of shortage and not one created by generic promotion. Grape shortages in Napa Valley, caused by replanting efforts to combat phylloxera and also by two poor vintages, occurred while the winery was in the midst of repositioning the Napa Valley brand further toward the high end of the wine market spectrum, placing greater emphasis on District and Reserve wines. With grape availability far below previous levels, grower prices jumped, requiring an important increase in wine prices. It also made it impossible for the Napa Valley brand to be in all the outlets where it formerly was sold. This was the opportune moment for the Coastal brand to develop a clear position in consumer minds and to fill the gap left by the Napa Valley brand, particularly in grocery and club stores.

BRAND OBJECTIVES

The task for the Napa Valley brand was to revitalize its image as the preeminent prestige wine and to manage the supply constraints without seriously alienating distributors and retailers accustomed to handling the brand. This was a difficult task because the shortage sometimes resulted in distributors having to pull Robert Mondavi Napa Valley from high-end stores. It took hard work to get those placements back once the wine had been pulled. Our efforts to manage the supply constraints while optimizing revenues and trade opportunities succeeded in establishing Napa Valley as the core brand of the portfolio.

BRAND STRATEGIES

A key consideration in our multibrand positioning effort was to build up the Napa Valley brand image so that it would benefit the Coastal and Woodbridge brands, which have a Robert Mondavi connection on the label. So we managed the supply constraints by shifting the prestige wines to the higher-priced, smaller production categories, resulting in more District and Reserve wines and less of the Napa Valley wines. During this period, the output of the high-end wines dropped from 550,000 cases to roughly 300,000 cases, reflecting the proportional drop in grape availability for these wines. By shifting the brand mix, elevating prices, and managing the shortage, we ended up enhancing Robert Mondavi's image and improving quality. This is evident, for example, in the high ratings and awards received by Robert Mondavi Chardonnay Reserve recently. This aura of high quality helped create a clear position for the Coastal brand as a wine of value. This brand hit the grocery store shelves right after the Napa Valley brand was pulled and at about the same price point.

BRAND TACTICS

The tactics for implementing the Napa Valley strategies included a greater emphasis on those restaurants and wine shops that were suitable outlets for higher-priced and lower-volume wines. The image of these outlets helped enhance Robert Mondavi's brand image, and vice versa. At the same time, the brand was moved away from grocery and club stores. Packaging, labeling, and pricing were all designed to enhance the brand image and perceived quality. The promotional mix changed as grocery store promotions, such as those used with Coastal and Woodbridge, were dropped and special promotion events such as wine maker dinners, awards, and tastings were emphasized. More special wines were introduced as we brought back a zinfandel after a long hiatus, introduced merlot, and released, experimentally, some California-produced Italian varietals. Robert Mondavi invested in creative-image advertising for the brand in several of the lifestyle and wine publications. This was, I believe, the first time that Robert Mondavi had undertaken an advertising campaign for the brand, so it was quite a leap for us.

ADVERTISING AND COMMUNICATIONS

The objectives of Robert Mondavi's advertising and communications plan were to enhance and clarify the overall image of Robert Mondavi, thereby positively affecting everything that carries the name; to reach as many premium wine consumers as possible; and to communicate quality, innovation, and leadership. The elements of the plan included a fully integrated program of advertising, promotion, and public relations devoted to the objectives. The program included print advertising in upscale magazines, a heavier-duty public relations campaign concerning what the winery was trying to accomplish, and direct marketing efforts to gauge consumer reactions. Consumer research undertaken prior to advertising revealed that Robert Mondavi had a comfortable but stodgy and conservative image that did not convey a sense of innovation or leading-edge work. There was clearly a need to change this image. Although the winery was trying to position itself at the price and quality levels of small boutique wineries and was using many of the same techniques and much of the same labor-intensive wine-making craft, the message was not getting through. This stimulated the launch of a public relations campaign to educate the trade and press about how Robert Mondavi's innovations and wine-making practices were producing wines with the same care and results as the boutiques.

The campaign worked; we did enhance and clarify the overall image of Robert Mondavi. The ads were very artistic, abstract, and unconventional in that they did not show a bottle of wine. This was completely different than anything else that had been done before. We reached as many premium wine consumers as possible in the selection of our advertising media, and we worked hard to communicate effectively the quality, innovation, and leadership associated with Robert Mondavi. Such an image-building campaign is designed with the future in mind and is often difficult to sell to marketing personnel, who are focused on immediate results. The goal was to enhance image rather than promote a specific product or sales drive, and this created some tensions in the allocation of budgets. At the time the image campaign was launched, the winery was hard-pressed to supply sufficient wines to meet demand, so it took a lot to convince people that image building was needed to ensure that increased quantities could be sold at higher prices in the future.

A complementary plan was started a couple of years after the basic Robert Mondavi image campaign was initiated. It called for separate brand-building advertising budgets for Coastal and Woodbridge. These brands were most closely associated with the Robert Mondavi name and benefited from the basic image plan, but they needed individual promotion efforts to clarify their positions. The elements of the program are brand-specific print advertising, spot radio promotions, and direct-trade education and promotion. Coastal has tended to focus more on radio spots, because they provide good support for local distribution and are more effective than national print. A one-minute spot seems effective in educating the consumer about the qualities of the Coastal brand. Woodbridge, on the other hand, is appealing to a wider audience of everyday wine users. It launched a new package and spent most of its advertising budget for a print campaign in lifestyle magazines. It is a different strategy seeking to obtain the same objectives as Coastal.

CONCLUSIONS

A fundamental question in the positioning of products is whether to extend existing brands to cover new products or markets, or to develop different brands. Robert Mondavi gave up the former strategy of using the Robert Mondavi brand for all products in favor of developing a stable of new brands, each one built around particular product and market characteristics. That choice has helped clarify the positions of the various brands for the sales force, distributors, restaurateurs, retailers, and consumers. The other side of this argument is that a brand extension policy would create confusion because buyers would less likely understand differences between the "extended" products and how they should be used. The positioning strategy must be executed by the sales force. Robert Mondavi relies heavily on the sales department and sales force to get the wines out there and into the right place so consumers can pull them through the distribution system. It is trying to clarify its messages by measuring consumer responses to them and then adjusting the messages so that they are better understood. Once the message is clarified, it is important that it is sent over a long enough period so that consumers really get the message and are able to act upon it. In the final analysis, the positioning of multiple wine brands requires common sense and adaptability, sound planning, thorough implementation, hard work, and good luck.

CHAPTER 21

Integrating the Marketing Elements

Kirby Moulton and James Lapsley

INTRODUCTION

This chapter examines how marketing elements are integrated in developing product strategies. The five elements that we are concerned with are product, packaging, promotion, price, and place. Just as the wine maker blends wines of different vineyards, varieties, and vintages to produce a preferred wine, the marketer blends strategies from the five elements to produce a preferred product position. Visualize a control panel with five dials, each controlling a set of relevant strategies, with an operator tweaking the dials to arrive at the proper mix for successful positioning. A creative operator can adjust the dials to achieve a synergy that gains the winery a competitive advantage. The key is how imaginative the operator is in reaching beyond the usual practices.

PRODUCT

A brand identifies a product or group of products as those of a manufacturer, producer, or distributor. Some wineries, such as Kendall-Jackson or Fetzer, use predominately one brand for all of its output. Others, such as E&J Gallo, create a series of brands for different qualities or types of products. Regardless of the strategy adopted, a good brand is worth money and can be sold, and the best brands have an almost personal relationship with the consumer that motivates repeat business. A brand's value derives from the image it conveys to the consumer about the quality and desirability of the branded product or products. Ideally the image helps differentiate the brand from competitors and is reinforced by repetition of message and by integrating the marketing mix to supply a consistent message to the consumer. For example, a brand may be positioned to be upscale and primarily seen in restaurants but not generally available on the retail shelf—thus positioning itself as a choice for a special occasion. Conversely, another brand may promote itself as consistently good tasting and thus be the solution for wine buyers overwhelmed by so many choices on the supermarket shelves. One of the jobs of the marketer is to help discern what the brand's unique personality might be, to identify where in the wine market such attributes add value, and then to craft a fully developed marketing plan consistent with the brand image.

Sustainable differentiation is a strategy for differentiating a product from its competitors on a long-term basis. A product can be differentiated in many ways, from capitalizing on the personality of the wine maker; to using particular production techniques, such as barrel fermentation or rotary fermenters; to using a particular grape variety. These strategies have been used in the past but generally have not provided a sustainable point of difference. Wine makers can be hired away, technology can be acquired by competitors, and grapes can be purchased. Ultimately, the vineyard may be the best means for differentiation for a winery, at least if the winery controls the vineyard. Location imparts quality characteristics to grapes that vary from region to region or vineyard to vineyard. Thus, a winery may choose to capitalize on the characteristics of the Santa Barbara, Monterey, Lodi, or Mendocino growing regions (in California), with the hope of finding the right price/quality relationship for positioning its brand. Similar comparisons can be made for growing regions in France, Spain, or Australia. Before positioning a product based on grapes from a specific vineyard, district, or region, the winery should assure itself that the grape supply is likely to be adequate to meet projected demand and that its product is in the correct market niche and can be economically competitive.

Once a brand has identified its points of differentiation and its place in the market, it also needs to identify, study, and learn from its competitors. Through use of market scanning information (if the winery is competing in supermarkets), or by talking with key retailers, a brand owner identifies a "competitive set" of wineries competing in the same price point or varietal/appellation. After identifying the competitors, the brand manager should attempt to define the competition's main points of differentiation, examine trends in package and bottle type, and review competitors' pricing and distributions strategies. And, of course, the winery must taste the competition.

Quality is a subjective, but real, factor and is a constantly moving target. Technology and increased grape quality due to better vineyard management have combined to increase the competition and narrow the difference in all price segments. The winery will need to conduct a large number of tastings to determine the particular quality characteristics of potential grape sources and to develop a blend and signature style for its own wines. For example, in the $7 to $10 subsector of the premium segment, the winery might blind-taste Woodbridge, Fetzer, Rosemount, Blackstone, Lindeman's, or R.H. Phillips and other competitive wines against its own products. The objective is to blend a wine that consumers will prefer to any of those in the blind tasting. The winery may use different varietals, barrel treatments, and fermentation processes to achieve such a result. The decision by professional wine tasters at the winery must subsequently be tested with consumers to validate or reject the blend. A wine's quality is based on the grapes from which it is made, the wine maker's art of producing and blending, and the judgment of consumers.

PACKAGING

As anyone walking down the wine aisle can deduce, packaging is an essential part of the marketing mix. One has only to observe the changes in labels and their positioning, bottle shapes, and logos that have occurred over the past decade to appreciate the competitive attention given to this marketing element. Obviously, design and color help create an image for the wine catch the consumer's eye. It is important that the label be consistent with the product. For instance, Barefoot Cellars, a competitor in the "house wine" section and destined to retail for $5.99, wants a label that is distinctive, easy to identify on the supermarket shelf, and suggests that wine is a fun, everyday beverage: The big blue footprint that is their logo accomplishes those goals. Conversely, an upscale wine probably will not require bold graphics that make the bottle stand out on a supermarket shelf or that demystifies wine, but instead will desire a more sophisticated label that looks at home in a upscale restaurant and invites the consumer to become involved with the imagery. The Steele Winery's label, with its oval shape, the involved Romanesque *S,* the faint pattern of shooting stars in the background, and the images of two screw heads on either side of the oval, as if the label were a wood plaque attached to the bottle by screws, all invite a second look and consumer involvement with the package. Interestingly, both label examples are effective in eliciting a positive emotional response from the consumer.

Simple and effective label designs are difficult to create, and their design should be left to experts in most cases. Good labels convey an image of producer integrity and the consumer's preferences. Designs can be extended to cartons and carrying cases by emphasizing the logo so that it might be seen clearly from a distance. A back label can offer a convincing story that attracts buyers to the wine. Label placement is important as well. Placing a label higher on the bottle than your competitors makes that bottle stand out on a shelf. Changes in bottle size or shape can be effective in building a desirable image, but the image may not be supported over time. The introduction of flange-topped bottles created a temporary competitive advantage for those wineries that adopted them. As more wineries used this "new" bottle, it began to lose its appeal as a way of differentiating one wine from another. Packaging can thus be used to help impart an image—for example, the use of a heavy bottle with a punt for an upscale wine—but since in most cases competitors will have access to the same glass manufacturer, the container is only one part of the packaging mix, helping to reinforce the overall marketing message.

PROMOTION

Promotion is essential to positioning a product in the mind of potential buyers. The best promotion results from creative thinking. Often a simple concept or design will be most effective. When R.H. Phillips was just beginning in the 1980s, and no one knew about the Dunnigan Hills, the winery commissioned a poster using its name and a striking image of a bird found in its vineyards. The image was tied to the winery's appellation and its label graphics. The poster was so stunning that it appeared in many restaurants and upscale hand-sell shops, thus helping promote the region while introducing the winery as a quality wine producer.

Giving away or even selling hats and T-shirts can be effective for spreading brand awareness. When Jed Steele, who was already well known from his days with Edmeades and Kendall-Jackson, started his own winery, he sent out to his friends in the trade a T-shirt with the image of two hands holding a cluster of grapes, a date, and the words "Birth of a Winery—Steele Wines." Although Jed did not have any wine to sell at that point, it reminded all of the wholesalers and retailers that Jed had a new winery, and, when they wore the shirt, they became promoters of his new venture and emotionally connected with the new winery, helping it to succeed. Wineries can also use hats, pins, and point-of-sale materials—shelf talkers, case talkers, presentation folders, and posters made from product shots—to spread their logo and to position their image.

Promotion that provides effective support for the winery's positioning objectives also motivates, distributors, retailers, and others in the marketing chain to buy into the same objectives. The creative use of imagery can be simple, such as large-format bottles with the brand name and logo for backbar placements in restaurants, or, in the case of retail sales, case printing so that a logo emerges when the cases are stacked together. A winery can launch this type of positioning support with a minimal advertising campaign using the design features adopted for labels and packaging. This might involve bold colors, easy-to-read graphics, and the logo, brand, and telephone number. This can be an effective format for an expensive full-page four-color ad in a top wine or fashion magazine, or for a simple flyer to the trade. Simplicity is usually the most effective. The objective is for consumers to remember the winery's colors, logos, and brand.

PRICE

Price is another key element in the marketing mix. Since in the long term price must have a relation to perceived quality, the price of a product says a lot about a winery's self-image. Wineries should approach pricing decisions in collaboration with distribution partners. Consideration must be given to the competitive situation, the bottle size, and the volume goals for the brand in the market. The pricing strategy is not restricted to a single price, but will often include a special price adjustment upon introduction of the product, a volume discount to be passed on to retailers, and a special feature price to be used for product promotions. The objective of the special price deals is to offer a lower price that will encourage new consumers to try the wine. The range of these special prices incorporates special situations that can be characterized as top-of-the-line, regular, special, and super special.

The top-of-the-line price position is for reserve wines or wines in short supply. The regular position is the price that consumers face day-to-day. The special price is for wines that are featured on special sales with margins that are almost always lower for the producer. The super special price is reserved for very specific sales, seeking fast turnover. Special and super special prices are often part of a larger marketing program that includes print ads and point-of-sale materials. These four pricing points might be as follows for a wine in the $7 to $14 premium category: top-of-the-line $11.99; regular $9.99; special $8.99; and super special $7.99.

PLACE

In developing its brand-marketing program, a winery must have specific distribution goals and be able to identify the type of stores appropriate for the brand. If a winery is a limited production operation located in a prestigious appellation, it may choose to sell almost all of its production direct from the winery, wholesaling to only select upper-end restaurants or retail accounts in order to sustain brand positioning. If, on the contrary, the winery has 500,000 cases of wine from California's Central Valley, the marketer will probably focus on supermarkets and chain stores (including drugstore and wine store chains). These outlets will be the major off-sale distribution objectives.

In working with wholesalers, the wine marketer needs to define clear sales objectives for a particular marketing plan. These objectives should specify the number and type of stores in which the brand seeks placement and the degree of penetration sought. One useful way to classify stores within a particular type is according to traffic volume in the stores (the number of customers per day). High-traffic stores will generally feature large wine displays at aisle ends or on islands elsewhere in the store. Medium-traffic stores may have two thirds the traffic of larger stores and have correspondingly smaller display opportunities. Low-traffic stores may have half or less of the volume of the large stores, and their floor displays may be only 12 to 15 cases as compared to 25 cases or more featured in high-traffic stores. A marketing plan may establish an objective to be in 85 percent of the high-traffic stores, 70 percent of the medium-traffic stores, and 50 percent of the low-traffic stores. The objectives may be for display in stores that will support the brand and/or wine category with extensive merchandising materials and feature advertising and that will offer advantageous shelf and cold-box positions at eye-level, adjacent to competition. The most effective distribution plans are those that are feasible and specific in identifying target stores and sales goals.

A pricing and discount arrangement will be needed to achieve brand-positioning objectives. Generally, these arrangements are based on depletion allowances tied to placement in targeted stores, or other sales incentives offered to wholesalers (in compliance with applicable laws in the specified market). Since discounts and incentives are expensive, the winery must be satisfied that the positioning results are worth it. They should be viewed as an investment that will effectively broaden distribution and, over time, build awareness and profits. For the marketing plan to work, it must be clear in its objectives. It must differentiate between an objective of deepening distribution through obtaining new accounts and/or better displays and floor stack, and an objective to enhance the brand's image by placing it in trend-setting restaurants. Each objective calls for a different product placement strategy.

The marketing plan should include strategies for gaining on-sale premise distribution if it is consistent with the winery's positioning objective. There are several types of on-sale opportunities. They include full bottle sales in restaurants, hotels, and other institutions; wine-the-glass programs in pubs, bars, lounges, and restaurants; and wine in special packages for airlines and transportation companies. Sometimes overlooked for on-premise sales are specific goals related to varietal by-the-glass promotions, wine list placements, and chain hotel banquet lists. The winery should explicitly examine the opportunities in these areas and then establish goals that are attainable. Wine-by-the-glass facilitates a brand introduction program because it allows buyers to listen to or read about seller recommendations, to consider the price, and to try the wine without having to buy a bottle.

Wine list strategies have changed over time as more and more on-premise outlets prepare their own lists using readily available computer programs. Consequently, the opportunity for a seller to furnish a complete wine list in return for listings on it have greatly diminished. Most sellers now must convince the buyer to list the wine based on its suitability for the buyer's wine program. Low price may not be the best strategy to gain a listing. For example, a wine selling to the on-sale premise for $55 a case might appear on the menu at $13.75, based on a three times markup. This may be the lowest priced wine on the list, and therefore not appealing to the majority of customers that appear to select wines for special occasions that call for something other than the least expensive choice. A wine-by-the-glass program can mitigate this reaction by offering the customer an opportunity to taste the wine before committing to buy a bottle. This may well attract more new consumers to a brand and move more wine than will wine list placement.

Hotels and caterers are often searching for less expensive wines with good brand recognition to offer as part of set dinners for conventions and other group activities. The seller needs to appraise carefully the likely clientele being served and then assess how the winery's brand can meet the client's needs. If the match is perceived to be good, it can result in substantial business volume. Value is the key in this situation, since it will provide an edge over competition at keeping banquet prices lower than they would otherwise be.

POSITIONING AND REPOSITIONING

Positioning and repositioning are essentially similar endeavors. They differ in that a new entrant to the market may lack brand recognition, whereas a repositioned brand often has some degree of brand recognition. The first step in either case is to develop a vision of the brand that encompasses where and how it will be used. This vision will need a story to go with it that communicates the vision to potential buyers and makes them at ease with the brand. It should identify the unique points of the brand and how they fit with market preferences. To be successful, the brand vision needs to be realistic and to respond to a large number of relevant market concerns.

Positioning and repositioning are dynamic processes because markets and consumers are continually changing. Market conditions are influenced by changes in market regulations, competitor strategies, distributor and retailer structure, and other similar factors. Consumer decisions are affected by changes in attitudes toward wine, occasions, weather, income, and many other subjective and objective factors. Some of these factors change rapidly and others change more slowly. The variability in consumer behavior and market conditions provides the opportunity for many different products to vie for buyer attention. Given these opportunities, a winery needs to continually evaluate the fit between the attributes of its product and the newly expressed preferences of the market. Knowledge of this link, or recognition that a gap exists between attributes and preferences provides the foundation for a successful repositioning strategy.

The winery's task is to analyze the brand attributes that have helped build the brand in the past and to determine if these attributes are still relevant in motivating consumer purchases of the repositioned brand. Changes in product, packaging, promotion, pricing, and/or distribution will be needed to develop new attributes more in tune with market needs. If the existing attributes are still relevant, they need to be emphasized to indicate how the brand is in touch with the changing market. The need for a continual evaluation of the attribute-preference link cannot be overemphasized. Even in the case of a new brand without a past history, the winery must ensure that the attribute-preference link and the related differentiation strategy used in planning are still relevant during execution of the marketing program.

There are many examples of brand repositioning carried out by Robert Mondavi, E&J Gallo, Beringer, and others that illustrate how the process works. The following example considers a wine brand in one market category that is recognized for having the best quality/price mix in the category. The brand owner wants to move the brand into the next-higher price category. The winery's vision is of a new wine that is reliable, affordable, and in touch with the tastes of more upscale consumers. The winery's research suggests that the most likely buyers will be those whose wine preferences and willingness to pay are both moving up. They may be shifting from, for example, from a Beringer white zinfandel at $4.99 or Lindeman's chardonnay at $5.99, to wines in the "fighting" varietal category, to wines in the $7.99 to $11.99 premium category with a higher-quality product image.

The $7.99 to $11.99 category is a major part of the rapidly growing premium-value varietal market. The major competitors in this segment include repositioned brands of some very large wineries such as E&J Gallo and Robert Mondavi. It also includes other producers that do not compete in the lower priced categories. Consequently, the premium value segment is less dominant by the very large wineries than is the fighting varietal segment. This means that the winery's brand will be associated more with super-premium brands than with mass marketed brands. The structure of the premium value segment and the presence of significantly differentiated wines allows a moderately sized winery to profit from the creative combination of its vineyard resources and wine making skills.

The most important competitors include Robert Mondavi Coastal, BV Coastal and Carneros, R. H. Phillips, Blackstone, and Gallo Sonoma. There are additional competitors, of course, and the list changes over time as competitors rise and fall.

Chardonnay, cabernet sauvignon, and merlot tend to be the most popular varietals in the target category. The example winery must decide whether to focus on one or more of these varieties or to focus on varieties more suitable for smaller market niches. The first strategy suggests dropping varieties such as French colombard, chenin blanc, and Napa gamay and concentrating on chardonnay, cabernet sauvignon and merlot from a specific appellation. This allows the winery to develop intensely competitive strategies focusing on a limited set of products. The second strategy, specializing in some less widely demanded varieties, allows the winery to position itself as *the* quality source for the markets preferring those varieties. In either case, a successful positioning strategy will be realistic, feasible, and based on an imaginative combination of marketing principles.

Scanning data can be used to test assumptions about competitive conditions. The data are obtained from supermarket wine sales, which account for 25 to 35 percent of total wine movement, depending on year and other circumstances. When carefully analyzed, these data can supply valuable information about competitive positions and responses. This will help guide positioning strategies and the evaluation of their effectiveness. Scanning data covers every one of the significant wine brands sold in supermarkets, and allows a winery to construct a scanning profile of its major competitors. The data do not cover movement in restaurants, independent retail establishments, or other institutional outlets, nor movement in states where supermarket sales of wines are restricted. The lessons are that scanning data do not characterize or describe the entire U.S. market and that conclusions based on this data should be carefully qualified. One source of bias occurs because the data are generated primarily in supermarkets, where sales are weighted toward wines selling for $10 or under, a market dominated primarily by large producers.

CONCLUSIONS

It has been said that you can only make a first impression once. But every time a wine brand is introduced or repositioned, it makes a new impression. Some brands, such as Vichon, have been repositioned three or four times in a decade. Each repositioning allowed it to present a new face but remain connected to a respected brand name. The challenge for smaller wineries is to creatively combine basic marketing principles while capitalizing on their own points of uniqueness, thus allowing them to compete successfully in markets dominated by large producers with greater resources. By combining the five *ps* of marketing (product, packaging, promotion, price, and place) with a coherent story that emphasizes product attributes in tune with a market segment, a small winery can succeed in today's market. Creativity, energy, and attention to the five *ps*, combined with a changing market, can make the smallest winery successful.

CHAPTER 22

Positioning: A Case Study

John Skupny

INTRODUCTION

This chapter is a case study about positioning, but it uses the example of creating a new brand that must be positioned in the market and ultimately in the consumer's mind. It discusses the use of a strategic audit to organize the elements of a successful brand. The audit includes identifying clearly the mission, objectives, and strategies to be followed. The strategies relate to deciding on product characteristics, sourcing grapes, finding custom production facilities, and determining packaging, distribution, pricing, and promotion tactics. Because this chapter focuses on marketing issues, it does not include other issues of concern, such as financing and internal organization. While considerable attention is devoted to product image, the key lesson is that image must be based on substance. As the experience of this winery indicates, improving a wine's reality will do wonders for its image.

Most of the ideas presented focus on image and image development in the ultra- and superpremium categories of wine, but the lessons of the chapter are also applicable to the other two categories, economy and popular premium.

John Skupny has worked for a number of California producers for the past 20 years, including Caymus Vineyards, Clos du Val Wine Company, and Niebaum-Coppola Estate Winery. In 1996 he and his family launched Lang & Reed Wine Company, which specializes in the production of Napa Valley cabernet franc.

SELECTING THE NAME

I had always thought that developing my own wine and designing my own label really would be fun. In actuality, it was the most torturous thing I've ever done. But a strategic audit helped. It provided a foundation for our business plan. My wife and I, with a couple of friends who had limited knowledge of the wine business, sat down and tried to conduct a strategic audit. The first consideration was the name. We had already agreed that the name of our enterprise was Lang and Reed Wine Company. The name reflects both the maternal side of my family and the maternal side of my wife's family. We did not think Skupny was an appropriate name to put on a wine label; it might be appropriate for beer or pickles, but definitely not wine.

The decision concerning name also reflected my experience in bringing many new wines to market for various employers. I learned that while an owner's name can be appealing, it is not necessarily so, and it cannot be forced on buyers. This is also true when a winery is sold and the sold winery carries an owner's name that is not compatible with the new buyer's brand. This is less of an issue if the name is more generic (as in Lang and Reed) than if it is unusual (as in Skupny).

THE STRATEGIC AUDIT

The objective of the strategic audit was to determine where the winery and its wines might be positioned. Our understanding of the market was that consumption was increasing in most premium wine categories and that varietal and site-specific designations were becoming very important. Thus, the winery strategy had to determine which varieties to focus on and what sort of designations to seek (California, North Coast, Napa Valley, Rutherford, or a specific vineyard). This, of course, leads to price positioning. The reality here is that a small winery such as we were establishing could not win at price competition. Our costs of production and distribution would not give us any competitive edge. Our prices would really be determined by the quality of production, as perceived by consumers, and the prices would not necessarily cover the costs of production if the consumer did not attach sufficient value to the wines.

The strategic analysis includes an analysis of competition. It revealed very daunting competitive conditions. There are excellent organizations with years of experience, skilled grape producers and wine makers, and effective selling programs. There are wineries and people making wine who have mature distribution channels, and there are people who own vineyards and wineries. I did not have these things, but I did have financial backing, 20 years of experience in the wine industry (as did my wife), and a defined market niche where we thought we could do well. We also had creative abilities honed in wine markets and strong links to the wine business.

TO OWN OR TO RENT

In assessing our financial resources, personal skills and experiences, and potential market needs, we decided to embark on our wine business venture without owning a vineyard or a winery. Our goal was to build a brand within a varietal niche, not an edifice. Once we established an image for the wine, we had to locate a source for consistently high-quality grapes and a facility where wines could be made to our specification and where sales growth would be possible. A long-term contract with a vineyard would permit the winery to set a style or shift to a different style. We had two people to help in making our wines but chose not to identify them on the label to avoid confusion about the identity of our own wine. We knew we might need to hire a consulting wine maker to facilitate production criteria and to lend his or her name to the project.

To date, all these decisions have proven to be good ones. They allowed us to concentrate on developing and marketing a quality wine without the headaches of ownership and maintenance. We are better off paying for those services than undertaking them ourselves.

THE PRODUCT AND ITS TESTING

The niche we selected is for cabernet franc varietal wines. We believed that we could create a unique wine with a positive image for a wine variety that was relatively unfamiliar to many consumers. In 1993, we made a prototype run of just 120 cases of our cabernet franc that was released in 1997. This was an aged wine, relative to our intended "early to market" wine and was basically a calling card to the trade to say "We've put our shingle out." The wine was very good, although if it had been in the normal cycle of what we are producing now, it should have been released in the fall of 1994. It was not. It was held back. We gave it just a little bit better wood than we had originally planned. It was made by a fairly famous pinot noir producer in California, so I think he handled it in the way I wanted it. He also worked for some of the most high-powered cabernet producers in Napa Valley, so he knew how to achieve a slightly lighter presentation wine with intensity and complexity.

Market testing provides important information about how a wine can be positioned. Obviously, the tested wine should be identical or close to identical to the marketed wine. In our case, circumstances led us to make a risky decision concerning the market testing of our wine. The original intention was to release the prototype in 1994 as an example of what the young wine would taste like, and then to produce wine from the 1994 vintage for release in the following year. In that way the marketed wine would be as similar as we could get to the tested wine. As it turned out, we delayed releasing the prototype because work commitments at my job made it impossible to produce wine each year after 1993. We resumed production in 1996 and also introduced our 1993 prototype to test the market. By that time I had left my job to devote all my time to developing our new company. We felt that the 1993 prototype really showed off the style that distinguished us as a producer of cabernet franc. Although it had more age than the wine we released in 1997 from the 1996 vintage, it was close enough to be a good predictor of what interesting things we could do with this variety.

We increased our output to about 1,200 cases in 1996, and about two-thirds of this was marketed in the fall of 1997, with the remainder released in late 1998, after 22 months of aging to give it the characteristics found in the 1993 vintage. The aged wine, called Premier Étage, still retained the Lang and Reed name but had a more sophisticated, upscale package. Our five-year objective is to raise output to about 4,000 cases. This is a risky decision that could go wrong for us, but it seems like a good gamble at this point.

We have tried to reduce our risk exposure through making high-quality wine and sensible marketing decisions. Developing a strong image for the winery and its products is also important in reducing market risks. We develop the image through public relations and other promotional activities.

WINERY OBJECTIVES

Our objectives are to make and sell wine and support our family. This is a family operation. That does not mean that nonfamily operations are doomed to failure; it is just that the family image is a powerful one in the marketplace. To accomplish our objectives, we will keep all aspects of the wine making as tightly controlled as possible. Such control is needed to get the feedback from our customers that permits us to make adjustments or otherwise alter our destiny. We do not have a large margin of error and strive to be nimble and flexible. If we become successful with cabernet franc and there is an opportunity elsewhere, we want to be able to shift. If suddenly the Food and Drug Administration says that cabernet franc causes facial blemishes and nobody wants to buy it, then we will have to be nimble enough to shift our brand image to another type of wine.

The strategic audit calls for the development of a mission statement to guide policies, strategies, and tactics. Our mission is to create, develop, and operate a small family-run wine company to produce high-quality wines that are enjoyable. The second part of our mission statement is to build a business that provides long-term family equity while improving the quality of life of our family and those who work with us. It is an enjoyable business if it is made to be that way.

PRODUCT IMAGE

To create enjoyable wine, we have chosen to make cabernet franc in a slightly different style than that generally commercially available from California wineries. I have found that most people making cabernet franc give it a claret or bordelaise character, which makes sense because this is an important variety in the Bordeaux region. On the other hand, there have been very few people who chose the lighter Loire or Beaujolais style for making cabernet franc wine. I say that I am making this wine in a "lighter" style even though I have trouble defining the word. It is like saying pinot noir is lighter than cabernet sauvignon, which it may be in appearance and image but not in taste. I have had pinot noir wines that were robust in flavor and as deep and long-lasting as any cabernet. It is just that the popular image is that one is lighter than the other, and I am banking on that image for my cabernet franc.

The favorable image arises from producing a limited quantity of cabernet franc wine from Napa Valley grapes and accenting the variety's characteristic aroma and fruity flavors. The term "fruity flavors" conveys the image of being easy to drink or of being "ultimately gulpable," using the phrase of a wine maker friend, as opposed to a wine that has more saturated flavors and tannin—a wine that requires bottle aging.

We have developed strategies in seven critical areas: organizing and managing, sourcing grapes, finding production facilities, selecting packaging, determining the means of distribution, establishing promotion and communication, and setting price levels. Is it not interesting that most of these strategies directly relate to imaging? They influence how consumers will position the product in their minds and how retailers will position it in their stores. We buy grapes from five or six different vineyards, some from the mountains, some from the valley, and some from the bench land. From an image standpoint, this means taking five or six different stories and trying to make an image that is greater than its parts. It is a different concept from building an image for a single vineyard. There are advantages and disadvantages to both approaches.

PRICING

The target retail price for the initial market entry was $18 per bottle. In addition, we later introduced a smaller quantity of "reserve" wine that is a little heavier and more tannic and is held back for extended barrel aging. We can make significantly more money at $25 to $30 a bottle than at $18. Actually, $18 is low for wines produced from Napa Valley grapes in small quantities, so the extra margin earned through the reserve wines balances out the leaner margins of the $18 wines. This strategy allows us to offer a great value in a ready-to-drink wine that is associated with a very high quality prestige wine of the same variety but with a higher price.

DISTRIBUTION

In positioning a wine, it is essential to develop a strategic alliance with brokers and distributors that understand the purpose and attributes of the business relationship. The alliances are entirely different for a winery that has 5,000 cases and a winery that has 150,000 cases. In organizing our alliances, I considered all the brokers and distributors that I worked with previously in my career and evaluated the likelihood that we could work together profitably. I interviewed many possibilities and was up front in identifying what our goals would be and how I thought we could work together. If I did not see a fit, I would tell the prospective distributor that.

My strategy was to sell the distributor on the value of our wine, including its profitability to the distributor, and then explain the marketing objectives, such as being in 10 intensive markets rather than 40 extensive markets. I laid out our expectations and constraints (for example, on travel and promotion) and then worked them into the distributor's expectations (this took a little negotiation) and involved the broker or distributor in deciding whether this was a good product and marketing idea. The goal was to arrive at a mutually attractive arrangement that would generate enthusiasm on both our parts.

We positioned the initial release as an early-to-market red wine with bright, fresh flavors and moderate tannin. This clearly distinguished it from wines made for the next millennium. We targeted it for distribution in California, New York, and London. In California, we planned to distribute directly for the first four or five years and then explore the use of brokers and distributors, following the strategies outlined above. We utilized a fine-wine distributor in New York and London from the beginning.

Our objective was to establish initial business in influential retail and restaurant accounts. This has created a high-quality image for the wine in the minds of consumers and the trade. We had to get maximum dollar on this early-release wine because we made such a small quantity. By selling the wine to high-prestige restaurants such as Spago, Boulevard, and Mustard's, we might make the next year's market more eager for the wine. Prospective new accounts might wonder "What did these guys know that I didn't? And how could they get the highest level of price for it?" This is how we developed the image that facilitated our growth.

It is important to maintain direct contact with all trade buyers to help in future growth and development. For example, I have heard wineries say something like this: "When we first came out, we used to sell 10 cases every year to Wally's down in West LA. Now that we've expanded our production to 10,000 cases, we find that Wally's bought only one case last year. What's the problem?" Well, the problem was one of two things. Either the wine was not hot anymore, which is a natural cycle of any winery or any product, or the winery lost Wally's as a client through neglect or through poor marketing and communications. The lesson that we learned is not to spread the sales effort too thin but to keep it focused and to keep in contact with our customers.

Our initial distribution was targeted to be about 65 percent restaurant trade, but when we reach full production, we expect to be at around 50 percent. This is an early-release red wine, and it needs to get to the restaurants promptly. Their advantage will be a quick turnover, a fast return on their money. It is best that all of a bright, fresh, vibrant wine be consumed in the first six months.

CONCLUSIONS

I've presented this case study of Lang and Reed to illustrate some important principles in developing and positioning a new brand. The first principle is that such development and positioning cannot take place outside of the total winery context. It is related to the financial resources available, the ability to find high-quality raw product and adequate production facilities, and the skills and interest level of winery personnel. Second, a strategic audit is an effective way of organizing information that is essential to being effective in the market. It identifies areas where decisions must be made before selecting a new product. The third principle is that product image is the result of several decisions concerning product style, brand, price, distribution outlets, packaging, and promotion. Finally, I believe that the case study supports the principle that a winery has to develop a favorable individual image that is expandable and unique in order to compete in an intensely competitive marketplace. Brand building is really a subset of imaging that conveys what the brand looks like, how the brand is to be presented, and what kind of feeling people should have about the brand.

PART V

Using Distribution

CHAPTER 23

Elements of a Wine Distribution Agreement

James M. Seff

INTRODUCTION

This chapter concerns the legal details of distribution arrangements. These are the details that are needed to implement effective distribution strategies. Other chapters in this book present the marketing considerations that underlie decisions about distribution. These considerations make sense to most marketing people; it is the legal details that are difficult to comprehend. Consequently, this chapter may be difficult for or less interesting to those with marketing backgrounds. However, if they pass it by, they risk endangering what are otherwise sensible distribution strategies. While the chapter applies specifically to the U.S. market, its ideas about attention to the legal environment surrounding distribution arrangements are applicable throughout the world.

Not every lawyer likes to draft contracts, just as, in medicine, brain surgery is not for everyone. At its best, documenting a complicated agreement appeals to lawyers who like to solve problems and do jigsaw puzzles. But in order to solve a problem, a person needs to understand it, and in order to write an effective agreement, a lawyer needs to understand the business purposes behind it.

James M. Seff heads the Wine, Beer & Spirits Law Group at Pillsbury Winthrop LLP (formerly Pillsbury Madison & Sutro LLP) in San Francisco. He has practiced wine law since 1969, has served as chief counsel of the Wine Institute and chair of the American Bar Association's Alcohol Beverage Law Committee, and is a professional member of the American Society for Enology and Viticulture.

The contract principles this chapter addresses are based on the common law of the United States, but they are also specific to the requirements of one or more of the individual states. U.S. contracts for the sale of wine or other alcoholic beverages are subject to (sometimes dramatically) different laws in each state. This creates many anomalous situations and a significant challenge for someone drafting an agreement for general utilization in each of the states.

THE LEGAL BACKGROUND

A bit of U.S. constitutional background may be useful. In order to avoid the economic impediments that had crippled much of Europe during the 18th century and that proved similarly destructive to the original 13 American colonies, the founding fathers incorporated the Commerce Clause into the Constitution.[1] The Commerce Clause prohibits one state from imposing an economic impediment to the free flow of commerce and goods from other states. Thus, under the Constitution, laws restricting the free flow of wine between the states should be illegal.

But, as a political price of the repeal of Prohibition in 1934, Congress adopted the 21st Amendment to the Constitution, the relevant section of which prohibits the transportation or importation of "intoxicating liquors" into a state in violation of that state's laws.[2] Under the 21st Amendment, as the U.S. Supreme Court has interpreted it over the years, states have broad rights to restrict various aspects of the trade in wine and other licensed beverages. So, for example, some states have strict controls on licensed beverage advertising and promotion, and each imposes a different excise tax and diverse rules about who can buy and sell licensed beverages. (Excise taxes are typically based on alcohol content and measured on a gallonage basis, although some states impose an ad valorem tax instead.) Those who intend to enter the U.S. wine trade on a national level thus face a daunting task. The only thing these laws have in common is their dissimilarity; in a word, the law is a mess.

Eighteen of the United States (Alabama, Idaho, Iowa, Maine, Michigan, Mississippi, Montana, New Hampshire, North Carolina, Ohio, Oregon, Pennsylvania, Utah, Vermont, Virginia, Washington, West Virginia, and Wyoming) and one important urban county (Montgomery County, Maryland), known as "control" or "monopoly" states, themselves actively participate in some aspects of the licensed beverage business. The State of

Wyoming is the sole and single wholesaler in that state but does not involve itself with the sale of wine at retail. The State of Pennsylvania is almost a "pure" control state; it buys all wine from out-of-state suppliers, sells it to restaurants for resale, and operates its own exclusive off-premise shops. The Pennsylvania Liquor Control Board is, in fact, the largest purchaser of licensed beverages in the United States. Of course, when states have monopolies on retail sales, there is no competition, and most members of the wine and spirits industry believe that state monopolies serve the industry and its consumers less well than private competitors in a free enterprise system.

But even among the majority of the "open" (i.e., not control) states, there is a bewildering lack of legal uniformity. For purposes of this chapter, however, the open states can be divided into two further subcategories: "franchise" and "good cause" states.

Good cause states are those that permit a supplier to fire a local wholesaler for general commercial reasons (usually nonperformance reasons) or, in some cases, for no specific reason at all (an "at will" termination). Franchise states are those that have adopted laws restricting a supplier's right to terminate a wholesaler to specific, limited reasons. A wine wholesaler typically does not enjoy what is usually thought of as a "franchise": an agreement with a supplier of products or services to purchase those products or services and market them under the supplier's name, for which the franchisee pays the franchisor a fee. Those more common franchise agreements are governed by different laws.

Laws restricting a supplier's right to terminate a wholesaler to specific, limited reasons are, of course, sponsored by the wholesalers themselves and adopted by legislatures that are traditionally closer to the local wholesaler than the out-of-state supplier. Franchise laws come in many different forms. Some create a franchise automatically when a supplier first ships wine into the state. Thus, an unwitting supplier may fill a small preliminary order with the intent to try the wholesaler out, only to find that it is now tied to the wholesale house for the foreseeable future.

Most franchise laws do permit a supplier to terminate a wholesaler for "good cause," but although the words may be the same, some states may interpret "good cause" in ways that are astonishingly restrictive. In Georgia, for example, the regulation makes a wholesaler's failure to maintain sales volume reasonably consistent with sales volumes of other wholesalers, or a wholesaler's failure to promote the product effectively, "good causes" for terminating or changing distributors.[3] A supplier that wants to change wholesalers must submit a detailed explanation of the specific business reasons for such changes to the State Revenue Commissioner (the wholesaler gets a copy at the same time, so it can file objections), who hears and determines the facts and makes the decision. Not surprisingly, the commissioner has historically favored local wholesalers over out-of-state suppliers in such disputes. In fact, the only truly effective way for a supplier to disengage from a Georgia wholesaler is to convince the wholesaler to give up the line voluntarily (i.e., to trade or sell it to another wholesaler). So drafting a distribution agreement for general application presents a real challenge.

THE IDEAL WINE DISTRIBUTION CONTRACT

Many clients who have been doing business with wholesalers on a handshake basis for years are not even aware that they have a binding oral contract. Most will admit that they prefer it that way, at least until they get into a serious dispute with the wholesaler. Then they may suddenly see the value of a written contract. Even today, most wineries probably do not have written distribution agreements.

The "ideal" wine distribution contract will vary among wineries and situations. If a winery has an immensely powerful brand, one sought after by competing wholesalers, it can, of course, call the shots. In that happy but unlikely situation, the winery's lawyer may want to draft an agreement that identifies and deals with as many potential points of friction in the ongoing relationship as possible. But even the most powerful brand owner will have some difficulty imposing a new contract (especially a demanding contract) on its existing wholesalers.

And even if the brand is one of the most powerful in the United States, it probably does not make sense to supply the same 15-page single-spaced document to the wholesaler in, say, Montana, as to the one in, say, Chicago. Depending on the contract's complexity and its renewal mechanisms, wholesalers in small states, or those that are in larger states but are doing relatively little business, simply do not provide the same business risks as do many major urban wholesalers. So it makes little sense to roll out the cannon where the laser will work better and be easier for the winery to administer.

In fact, few wineries decide to interrupt ongoing relationships by offering existing wholesalers written contracts (if they do not have them already). And new entrants to a market typically lack the economic power to demand concessions from their wholesalers. Indeed, there are so many brands competing for so little shelf space that a new winery is lucky to find a wholesaler at all. In a typical year, the federal Bureau of Alcohol, Tobacco, and Firearms examines approximately 60,000 licensed beverage labels offered for sale in the United States and rejects about 20 percent of them.[4] And in California, the number of wine wholesalers has shrunk from over 50 about 30 years ago to about 12 today. And only 2 of these are large, multibrand, full-service wine and spirits wholesalers. California's experience is becoming the U.S. norm, rather than the exception. There are more than 2,100 wineries in the United States today, while about 500 wholesalers—most of them small but a few of them enormous—control 90 percent of the market. So for the new, small winery, or the winery that has not previously traded in a given state, a short contract may be all it needs or all it can get.

Contracts naturally reflect the comparative economic bargaining power of the parties. Having written and reviewed a great many such contracts, I have concluded that, from the winery's point of view, they serve one principal purpose: to make it as easy as possible for a winery to fire its wholesaler when the wholesaler has failed to live up to its part of the distribution bargain. And yet in some ways the wholesaler has the natural advantage. It can effectively terminate the relationship by simply not reordering. In most cases, a distributor is unlikely to care if a small winery wishes to go elsewhere. But as the dollar value of the winery's business increases, so will the wholesaler's investment in time, money, and effort (not to mention ego, which is often a major factor). The more money the wholesaler makes on the wine, and the higher the wine's prestige value to the wholesaler's business, the harder it will fight to keep the line.

Rather than trudge through a distribution agreement clause by clause, I will use my firm's long-form agreement as an outline and note only some of the more interesting or difficult issues each section presents.

RECITAL

If a recital is used in the contract, it should include the consideration that underlies the contract. Usually, the mutual covenants and agreements contained in the contract are sufficient consideration, particularly if they are acknowledged as such in the recital.

APPOINTMENT

This is what the wholesaler has bargained for, especially if it will be the exclusive wholesaler in the territory. In such a case, the Uniform Commercial Code requires the wholesaler to use its best efforts to promote and sell the line,[5] a greater commitment and obligation on the wholesaler's part than it might realize. The winery may wish to make the appointment apply to line extensions, or to specify the brands for which the wholesaler is responsible. It is not uncommon for a supplier of many brands to split them between different houses in the same market.

TERRITORY RESTRICTIONS

Most wholesalers will only accept an exclusive appointment for a particular territory (i.e., a geographic region or a customer subset) except in the few states that make exclusive territories illegal. Sometimes, a new wholesaler will accept a dual appointment for a short time, but dual wholesalers often fail to work a brand because the other wholesaler might get the benefit. Formerly, area of principal responsibility (APR) clauses specified an understanding that the APR did not place "a territorial or customer restriction" on distributors' sales (unless required by state law) and that the designation of the APR was solely to judge the wholesaler's performance under the agreement. This language was necessary because the antitrust laws used to prohibit certain kinds of vertical restrictions. But the laws have substantially relaxed[6] and most vertical restrictions (other than restraints on resale prices) are judged by a "rule of reason" that looks to the winery's market power and whether the restriction is anti- or procompetition. In the United States, no winery, with the possible exception of Gallo, needs to be too worried about imposing vertical territorial restrictions on wholesalers to ensure that the wholesalers will aggressively market the wines.

Unfortunately, the winery may find itself in the unwelcome role of police officer if one wholesaler encroaches on another's exclusive territory. The winery may expose itself to certain antitrust claims in attempting to adjudicate competing claims of exclusive wholesalers. The key point is that the winery must impose and administer territorial restrictions for its own benefit to increase sales vis-à-vis other wineries—and not as the agent for wholesalers acting in concert to divide markets among themselves.

TERMS OF SALE

The distribution agreement should not specify price terms, because the winery wants to retain maximum flexibility to change prices. Instead, it should indicate that the wholesaler must pay the invoice price. The winery should have the right to raise its price at any time, although many contracts require at least 30 days' advance notice. The contract should make the wholesaler's order subject to acceptance by the winery at its principal office and should specify that all outstanding orders are cancelled if the agreement is terminated. If the order is subject to acceptance at the winery, the winery need not fill it, even if the contract does not provide other protection elsewhere. The winery should retain the right to determine and adjust credit limits for the wholesaler. Occasionally, a credit clause will require the wholesaler to send the winery current profit and loss statements and balance sheets.

Most winery sales are priced at the winery, with the buyer accepting ownership at that point. A long-form contract will often establish that the common carrier is the wholesaler's agent. This passes the risk of any loss during transit to the wholesaler when the carrier picks up the wines, even if the winery selects the carrier. In cases where a state requires the payment of use taxes on goods used (but not purchased) in the state, such as pallets, point-of-sale material, and packaging, the agreement may stipulate that the payment of such use taxes is the wholesaler's responsibility.

PERFORMANCE STANDARDS

The vital part of this clause is the distributor's agreement to use its best efforts to meet or exceed purchase, sale, and breadth of distribution requirements (i.e., placements of wine in the required number of on-sale and off-sale accounts by quality of account). If the wholesaler is exclusive, the Uniform Commercial Code implies a best efforts requirement. The wholesaler's failure to achieve its agreed-upon goals should be good cause for termination, even in many franchise states. However, for termination to be supported, the distributor's obligation should be documented. The winery would prefer to have the wholesaler's unqualified agreement to achieve the specified goals and objectives but cannot reasonably expect to get this agreement. A "best efforts" requirement, however, is almost as good, because in most states it imposes a high standard of performance.

A winery may also require a wholesaler to maintain an adequate sales force, to cooperate with the winery's marketing representatives and sales-people in its APR, to maintain adequate inventory levels (e.g., to cover at least 30 days' anticipated sales), to furnish depletion and other reports on a periodic basis, to let the winery take a physical inventory of the winery's products and point-of-sale material at the wholesaler's location, to report market trends and competitive activities in the APR, to observe all applicable laws and regulations, and to indemnify the winery against any loss from breaking those laws. The wholesaler should also agree to use its best efforts to sell only wine of merchantable quality and to return damaged goods (or destroy them, with the winery's permission), to maintain reasonable storage standards (including temperature and stock rotation guidelines), and to use the winery's trademarks only for specified purposes and only so long as the distribution agreement exists.

With the rising importance of intellectual property, some distribution agreements go into substantial detail about the wholesaler's use of the supplier's trademarks, and many impose on the wholesaler an obligation to notify the winery of any trademark infringement or unfair competition it learns about. Some even require the wholesaler to assist the winery in defending the winery's trademarks (but at the winery's own expense).

The supplier and wholesaler may both agree that the winery selected the wholesaler based on the wholesaler's representations that it could do the job. If the wholesaler misrepresented its capacities, many courts would find that misrepresentation a material breach of contract that would permit termination.

Older contracts provided a long and elegant mechanism for establishing annual sales and depletion goals and objectives. Some wholesalers may demand such a mechanism, but experience in the last several years has been that most will simply agree in advance to whatever annual goals and objectives the winery sets. This gives the winery substantial power, which, of course, it should use reasonably and fairly. Courts are unlikely to enforce goals and objectives that are patently unreasonable or that the winery appears to have established for the purpose of making the wholesaler fail. A wholesaler's failure to reach reasonable goals and objectives is probably the most common reason for a termination.

TERMS OF AGREEMENT

Most of my firm's distribution agreements are now evergreen agreements, which renew automatically unless terminated. Formerly, the wholesaler was required to meet or exceed its annual goals and objectives in order to get the automatic renewal. If the wholesaler missed its numbers, my firm provided an opportunity for both parties to sign a written extension 30 days before the agreement expired. This is orderly and thorough but ignores the practicalities of the wine business. Even most larger wineries do not have the staff or the inclination to administer their distribution agreements with such precision. In most cases, a winery keeps shipping to its wholesaler even if the wholesaler misses its numbers. Wineries seldom go to the trouble of a written renewal. My firm believes it is better to permit the contract to continue automatically, partly in recognition of commercial reality, but also to avoid a claim by a terminated wholesaler that it was no longer governed by a contract it signed some years ago but that, by its terms, had expired. But my firm reserves the winery's right to terminate the contract for nonperformance.

TERMINATION

The termination clause should be guided by a practical axiom: Whenever possible, a winery should terminate a wholesaler only when the wholesaler does not owe the winery much, or any, money. Once a winery cuts the cord, especially when the wholesaler will lose substantial revenue from the loss of the brand, it should be no surprise that the wholesaler will no longer be the winery's best friend. In most disputed terminations, the wholesaler's immediate reaction will be to withhold any money it owes the winery, and force the winery into either suing for it or granting at least part of it as the winery's cost to get out of the deal. Thus, the timing of the termination (where the winery has the luxury to plan in advance) can be critical and can have an immediate, measurable payback.

When a winery does have the luxury to plan (as it usually will), it should carefully document its dissatisfaction with the wholesaler and build an appropriate record prior to termination. The complaints need not be mean-spirited, but they should be direct: "We are frankly disappointed and dissatisfied with your poor performance in on-premise accounts during the last quarter. Please remember your agreement to place 35 percent of our wines in on-premise accounts by the end of the contract year." Indeed, the winery should begin building a termination file, even when things are going well. The winery should, by all means, compliment and encourage its wholesaler for good work, but congratulatory letters have a way of coming back to bite if the relationship goes sour. This follows another axiom: Compliment a wholesaler orally; complain in writing.

A court forced to adjudicate an unlawful termination claim will consider the period of time the relationship has existed; the amount of wine (from the winery and from competing wineries) the wholesaler has sold and is now selling; the trends in the wholesaler's purchases and depletions (compared to local, regional, or national trends for the same wine and the industry in general); the wholesaler's investment in the brand, in terms of money, effort, and promotion (the wholesaler will nearly always overstate the investment, and the winery will usually tend to minimize it); and like factors. Of course, the law in the wholesaler's state will usually control. In California and other jurisdictions, industry practice can be a critical issue.

In terms of the distribution agreement itself, the winery is always best off if it can terminate the agreement without cause on, say, 30 days' notice. Although the wholesaler does not need parallel protection, because it can terminate the agreement de facto simply by no longer making purchases, it may still be useful to make the termination clause bilateral.

In any case, the winery should have the right to terminate immediately, or at least on short notice, based on the wholesaler's insolvency, business suspension, the rendition of unsatisfied judgments against the wholesaler, the issuance of a voluntary or involuntary lien against the wholesaler's property, and similar economic reasons. The winery should absolutely insist upon the right to terminate the contract on a change of the wholesaler's ownership or active management (the winery might also want to terminate if it is acquired by, or merges with, another winery with a stronger distribution system), if the wholesaler defaults in the payment of money it owes the winery, and for the wholesaler's failure to perform any of its material contractual obligations. All of these grounds can be made reciprocal, which frequently makes them easier to sell to the wholesaler.

My firm's long-form contract contains numerous other grounds for winery termination. Experience shows that if a new wholesaler wants the brand badly enough, it will accept those grounds.

LIQUIDATED DAMAGES

A properly drafted liquidated damages clause can be worth serious money if a court rules a winery terminated its wholesaler unlawfully. Recently, a large California winery with a powerful brand terminated one of its long-term wholesalers. The wholesaler claimed that the termination was unlawful and that the loss of the brand would cost it millions of dollars. The winery asserted a liquidated damages clause as a defense. The clause was wordy—over a page of single-spaced typing. The wholesaler argued that, under California law, the clause must be based on an actual economic study of the potential lost value. The court rejected the argument, holding that the clause in question was reasonable. The clause saved the winery millions.

We usually state liquidated damages as a per case amount for each case of wine the wholesaler depleted, less returns, during the 12 months preceding notice of termination. But other formulas, for example, a percentage of annual sales, work just as well. In any case, it can be critical to state clearly the period that will serve as the measure of the liquidated damages amount. And even when this is done, a court or arbitrator may require proof that it is reasonable.

EVENTS FOLLOWING EXPIRATION OR TERMINATION

The winery should protect itself if and when the contract ends. Following are some of the basic ways:

- Consider that all of the wholesaler's outstanding unshipped orders are canceled.
- Only honor the wholesaler's outstanding orders for its historical requirements during the period between the notice of termination and its effective date. This is to prevent the wholesaler, out of spite or a desire to make one final killing on the brand, from selling it out at very low prices that could injure or, in extreme cases, even destroy, the brand's goodwill.

- Place the wholesaler on a cash basis for any further orders.
- Provide that the winery, or its designee (typically the new wholesaler), can, at its option, buy wines in good and salable condition from the old wholesaler's inventory, along with advertising and other point-of-sale material.

The wholesaler should agree to return or sell back other promotional material at cost and to remove any of the winery's trademarks from its stationery, business cards, trucks, and the like.

Both parties should warrant and covenant that, after termination, neither will take any action to diminish the other's business or goodwill.

CLAIMS, DAMAGES AND WAIVER

In the long-form contract, my firm likes to require the wholesaler to inform the winery in writing of any complaints or claims the wholesaler has against the winery. The clause provides that failure to do so waives such claims.

WARRANTIES

Most state laws imply warranties of merchantability and fitness for a particular purpose. My firm typically warrants that a winery will produce, package, and label all products in compliance with law. The wholesaler usually gets the right to return for full credit all wines that became unmerchantable because of the winery's negligence, but my firm restricts the wholesaler's right to dispose of such wines in other ways. My firm usually disclaims the implied warranties in large type so there can be no subsequent argument about hiding the disclaimer in small print.

FORCE MAJEURE

Most contracts for the sale of goods and services contain a clause that excuses a party from nonperformance due to an event beyond its control, such as an act of God, fire, flood, strike, war, and (a favorite in California) earthquake. The exemption lasts for the duration of the unexpected event.

NOTICES

All notices should be in writing, and the contract should specify the appropriate means of forwarding, including certified or registered mail, facsimile transmissions, and e-mail. In each case, the contract should specify when the notice becomes effective, e.g., date and time of postmark, transmission, or signed receipt.

SEVERABILITY

This standard clause provides that if a term or provision of the contract violates the law, it shall be amended or deleted without changing any other term of the agreement. This clause is especially important in contracts for general use in every state, because local laws vary so much.

GOVERNING LAW

The choice of law and venue can be critical because parties have an obvious and understandable reluctance to litigate under somebody else's law and, especially, in the other party's home court. My firm always asks that California law apply (after all, the contract is made in California, and each order is valid only if accepted at the winery's home office in California) and lays venue in a California court in San Francisco, or elsewhere near the winery. The clause should require that both parties submit to that court's jurisdiction. This is also a good place to specify the use of alternative dispute resolution mechanisms. My firm has used arbitration effectively in the past, although its costs continue to rise, and mediation of many disputes (except where the wholesaler refuses to pay its bills) can be cost-effective and is becoming more common.

WAIVER

The standard waiver clause provides that failure by a party to insist upon or enforce any term of the contract is not a waiver of that term, unless in writing, signed by the party against whom the waiver is sought to be enforced. Likewise, waiver of any one term is not a waiver of any other, and waiver of a term on any one occasion shall not be a waiver of the same term on all occasions.

ADHERENCE TO LAWS

This is a nonstandard clause under which the wholesaler acknowledges that the winery's policy is to comply with all relevant laws and regulations (particularly antitrust laws, because wholesalers often assert them as additional claims in actions for unlawful termination), that any action by a winery employee contrary to this policy will not bind either party, and that the wholesaler will immediately give the winery written notice of any attempt by any of the winery's employees to violate winery policy. The theory, of course, is to permit the winery to argue that the wholesaler waived any claim it did not report. To my knowledge, this clause has never been tested in court.

ENTIRE AGREEMENT

This standard integration clause says the agreement supersedes prior oral and written agreements and can be modified only in writing. For extra measure, the lawyer can add an acknowledgment that the contract has been negotiated in good faith and at arm's length and has been entered into freely.

NATURE OF THE AGREEMENT

This is a standard but important clause that makes it clear that the wholesaler is not the winery's agent or franchisee and that neither party can act as if it were, or create any obligations on behalf of the other party.

BINDING AGREEMENT

This standard clause makes the contract binding, not only on the parties, but on their successors, a very important provision if the wholesaler comes under new management (to which the winery does not object) and the agreement is to be kept in effect. It can also be particularly important if the contract predates a nonretroactive franchise law.

CONCLUSIONS

As noted above, many (perhaps most) wineries have no formal written distribution agreements with their wholesalers. Where such agreements are in place, they are usually short and cover only the most critical issues. Many of the concepts on my firm's long-form agreement are already contained or implied in the state's general commercial law. Many of the provisions simply try to anticipate commercial difficulties my firm has run across in the past.

Lawyers should take care not to "overlawyer." Too much law can get in the way of a business deal, and the last thing one party to a deal wants to do is to frighten the other side away by lengthy, unreasonable (and, frankly, often unnecessary) legal requirements. On the other hand, for a big supplier with a valuable brand, using as much detail as possible may be both prudent and appropriate.

REFERENCES

1. U.S. CONST. art. I, § 8, cl. 3.
2. U.S. CONST. amend. XXI, § 2.
3. Official Compilation Rules and Regulations Georgia Regulations, Chap. 560–2–5–.02.
4. Yahoo Finance, "Trade with China? How about PNTR with Maryland?" May 25, 2000. Accessed in June.
5. See, e.g., CAL. COMM. CODE § 2306(b). See footnote 13 on page 7.
6. *Continental T.V., Inc., et al. v. GTE Sylvania Inc.*, 433 U.S. 36 (1977).

CHAPTER **24**

Making Distribution Work

Richard A. Gooner

INTRODUCTION

The chapter focuses on the basics that most marketing and general management practitioners can and should apply to improve the effectiveness of their distribution decisions. It concerns the choices of distribution partners and the manner of dealing with them. The underlying premise is that the most expensive and painful distribution problems come from inattention to business basics and a failure to plan explicitly, rather than from poor legal or financial counsel or similar "fine points."

In the current industry environment, no wine or spirits supplier—regardless of size, profitability, or organizational characteristics—can force distribution to work. Neither Bacardi nor Gallo exercises the level of control that it would like over its own distribution companies. Control of "independent" distribution by suppliers such as Brown-Forman and Glen Ellen is even less effective. Consolidation of distributors, a proliferation of brands, and the liberalization of global trade have shifted power in the distribution chain from producers to distributors.

Wineries pursue a range of strategies to address changes in the distribution system. The strategies may be grouped into three highly aggregated and subjective categories used for discussion in this chapter: (1) a denial of the shift in power in distribution and a unilateral and adversarial ap-

Richard A. Gooner received his PhD in marketing from the University of North Carolina and currently teaches at the University of Alabama. Prior to returning to academia, Richard was vice president of sales and marketing for Glen Ellen winery and was vice president of international brand development for Brown-Forman.

proach to dealing with distribution, (2) acquiescence and abrogation of all marketing and sales to distributors, and (3) a positive bilateral approach that seeks new ways to improve results for both distributors and producers. The third approach is the most rewarding strategically and financially. It provides both the supplier and the distributor the most control and satisfaction. It also involves the most work and is the most difficult to execute.

This chapter discusses the rationale of the bilateral approach and analyzes the building blocks that support it. It stresses the need for explicit planning, detailed knowledge of each business, and agreement on goals, tactics, and evaluation. It reviews important elements in establishing and maintaining distribution, including choosing the distributor, arriving at an agreement, working with the distributor, and terminating the relationship.

THE DISTRIBUTION SYSTEM

Two words come to mind in describing recent developments in the wine distribution system: instability and consolidation. Instability arose from the rapid change in wine and spirits production, the mix of products, consumer preferences, and the technology of marketing. Consolidation arose from instability, the availability of capital, and economies of scale in modern marketing. While wine distribution is still fairly fragmented in the United States, due largely to legal constraints, U.S. Census data indicate that the top 25 U.S. distributors account for about one-half of the U.S. market sales, double their market share 10 years earlier. The result is evident in California, where the ratio of wineries to distributors changed from about 9-to-1 to about 200-to-1 over a period of 20 years. A consolidation in distribution systems is evident in Europe and Asia as well. These changes force wineries to review their distribution policies. For some wineries, this review has revealed that they really do not have an explicit policy stating what they expect from a distributor. They have to rediscover how to choose a distributor.

SELECTING A DISTRIBUTION OPTION

The characteristics of the winery, its product, and the target market dictate where the winery should go for distribution. Some hard reflection and objective analysis should take place before a winery begins the distributor selection process. For example, before selecting a distribution strategy, a winery must identify what the product is and what position is being sought for it. It should determine whether available volumes will fit with potential channels and markets. It should assess whether its own financial and human resources can support intensive travel, training, and introduction activities. A realistic evaluation of market and business conditions as well as the firm's condition should drive strategies and tactics; manager ambitions or owner egos should not.

Most of the discussion presented in this chapter focuses on distribution through distributors. However, the principles involved apply, to varying degrees, to other distribution arrangements. In each case, there is a clear need for detailed and objective analysis prior to making a decision. A winery is limited by only its finances and imagination in tailoring its selling effort to individual markets. The caveat is that the winery must know what each entity will contribute to the final result and at what cost.

The alternative channels available for wine distribution include:

- direct sales from winery sales room and by mail order and the Internet
- direct sales by sales force selling to retailers, food service outlets, and distributors
- sales by winery sales force to distributors only
- sales through brokers only, with no winery sales force
- sales through a marketing company

Which alternative a winery chooses will depend on cost, winery production levels and mix, brand strength, pricing policy, competitive conditions, location of target markets, and location of production facilities.

Knowledge of the costs of different distribution options is essential. The firm's financial manager can help in this assessment by identifying relevant cost points and assisting in the analysis. While the firm may not obtain "financial statement" accuracy in estimating the numbers, it will learn a great deal about the implications of various distribution choices in different markets. It is particularly important to evaluate volume-cost relationships in order to support financing decisions and efforts to coordinate goals with selected distributors.

The following example illustrates how margins and costs might vary. It is intended to underscore the necessity for careful analysis before choosing a distribution method. A winery with a cost of production of $21.50 per case may be able to sell to a broker or marketing company for $25.50 per case and earn $4.00 per case in gross margin, with virtually no added marketing costs. However, the winery has no control over subsequent marketing, and volume may not meet expectations. The same winery might sell directly to a distributor for $30 per case and obtain a gross margin of $8.50. This will be needed to cover freight, promotion, and field sales costs, and the net margin may be no better than earned from a broker. On the other hand, the break-even point can be lower and the margins higher than with brokers if the volume is greater. The winery may have more control over its marketing, depending on sales agreements and how its sales force interacts with distributors. Another option is for the winery to sell to retailers (including restaurants) directly. The gross margin would be higher, for example, $18.50, but the winery will incur higher freight, overhead, and sales costs. Generally, a high-volume winery will find that the break-even price for direct sales is higher and the margin is lower than for sales through distributors. Selling directly to consumers, at $60 per case in this example, yields the greatest margin and control over marketing but also entails the highest costs (to cover facilities, mailing lists, personnel, and overhead). When wineries sell directly to consumers, the wineries will also be constrained by the volume that can be moved.

A planning framework used by a major producer of varietal wines details the elements that need analysis before choosing a distribution strategy. The producer first identifies key target markets, for example, New York, Texas, or London. It then determines the types of channels and outlets within the key markets. These might be supermarket chains, restaurants, and wine or spirits distributors. Finally, it identifies the key accounts within each channel, such as Safeway, Odd Bins, and Olive Garden. Much of this information has to be gathered through surveys and interviews in the field, and a search through trade publications and other sources of marketing information, including the Internet.

Once this part of the planning is completed, then the winery may have a better idea of exactly what it wants from its distribution system. It should be able to answer the following questions:

- What types of distributors should handle the product?
- What are the channel members' attitudes toward and motivations for handling the product?
- What intensity of wholesale and retail coverage is needed?
- What margins are appropriate?
- What forms of physical distribution are needed?

Once these questions are answered, the producer can list the characteristics of an ideal distributor. The key point is not the list itself but the thinking, tradeoffs, and choices that go into making the list. The winery should be tough on itself in deciding what the "must have" characteristics are. Experience suggests that a winery will be lucky to find most of its "musts" and many of its "importants" among the distributors being considered. The list will call for information on financial integrity, brands handled, trends in volume and product mix, management philosophy and competence, loyalty to suppliers and customers, and areas of expertise. This investigation is like the kind of "due diligence" effort expected in mergers and acquisitions. The sources of information are interviews with suppliers and customers in the trade, and surveys, credit reports, and observations in various market outlets.

It is important for the winery to be discreet when searching out distribution options so that existing distributors do not feel threatened. This is less of an issue if the winery is introducing a new line or opening a new market, or if a distributor is terminating the winery. With that caution in mind, one of the best sources for information is a winery's own sales representatives in the market. They should know who the distributors are, how effective their representation is, and what the trade thinks of them. The winery should identify competitive products and determine how they are merchandised and by whom.

DISTRIBUTOR POLICY AND AGREEMENTS

A distributor policy contains a clear and consistent articulation of the firm's philosophy. It outlines the winery's and distributor's expectations, responsibilities, and commitments. It is the umbrella under which a legal distributor agreement is negotiated. The winery and distributor should discuss and contribute to the policy statement, which should be kept short so that people can refer to it easily. It is a one- to three-page document that can be used by sales and marketing personnel to help them better understand what is expected from them and what support they can expect. This statement is not a legal document but should be reviewed by an attorney familiar with marketing, because it might be subpoenaed if litigation occurs.

The distributor agreement is a legal document requiring careful attention and precision in specifying expectations and commitments. Everything should be negotiable, and the agreement should tackle "tough" issues that could be subject to future misunderstanding. However, there is no way that one party can "force" the other to act contrary to its own interest. Coverage should include the term of the agreement, the functions performed by the distributor and those performed by the winery, pricing, payment and credit terms, delivery and shipment terms, inventory arrangements, dispute settlement procedures, grounds for termination, and other details of the relationship. The agreements should be for a fixed term with an agreed arrangement for renewal or cancellation. Even with cancellation clauses, an agreement should never be undertaken with the belief that it can be easily broken. If the agreement has a binding arbitration clause, which is generally a good idea because it saves legal expenses, it is to the winery's interest to have any disputes subject to binding arbitration in the winery's home state. Verbal commitments must be reflected in the agreement, or they may be difficult to enforce.

WORKING TOGETHER

The winery and the distributor are setting up a business-building arrangement in which both seek to make money. The elements of the working relationship should be negotiated before the final agreement is signed. These elements include performance standards, clear goals that meet most of each party's main objectives, measurement of results against these goals, communications, and joint working arrangements.

Performance standards define what the winery expects the distributor to do. Too often this area is ignored because each party assumes that the other knows what will happen. Examples of standards include pricing and payment procedures, the frequency of distributor contacts with customers, which clients are contacted, how point-of-sale materials are obtained and distributed, how the brand will be featured in advertising, the procedure for following up on winery referrals, and the responsibility for organizing wine maker dinners. The objective is to have a set of standards that makes it easy to assess objectively whether the distributor and the winery have met their respective commitments.

The best standards are specific, measurable, achievable, and consistent with each other and with the distributor's other commitments. They should be specific enough that a third party could readily understand what was expected of the supplier and the distributor. They should also be measurable, so that a third party could review the records and determine whether standards had been followed and objectives achieved. Achievable standards are those that employees can reasonably accomplish. For example, quotas that are set too high will discourage performance from the outset, rather than motivate effort.

Unless care is taken, some standards may be mutually exclusive. For example, can the distributor effectively secure new accounts (one standard) at the same time it is serving existing accounts (another standard)? Is it reasonable to expect the distributor to maximize the winery's return (one standard) while trying to maximize winery market share and growth (other standards)? To avoid incompatible standards, it is necessary to identify those that are most important and to work those into the agreement.

Setting goals is a cooperative effort. There is no point in a winery attempting to dictate them, because the distributor is likely to accept those that are too low and ignore those that are thought to be unreasonable. The preferred way is to get out into the market and learn its true potential, by talking with the trade and the distributor, and then discussing conclusions with the distributor to arrive at a set of mutually acceptable goals. By doing advance planning, taking time to study the market, working with salespeople, and trying to understand the business, the winery can narrow the areas of disagreement between the winery and distributor and make negotiation easier. This is a mutual market-based approach to setting goals that establishes the supplier and distributor as equal players. It is not an inexpensive activity, but it fixes in the distributor's mind that the winery is "engaged" in the negotiation process and cannot be easily ignored.

Most suppliers burden distributors with unilateral and unrealistic quotas. It is easier for the distributors to accede, never intending to achieve the full goal. The suppliers do not recognize that they have not engaged the distributor in meaningful dialogue, and they go away happy for a few weeks or months. The mutual market-based approach—utilizing the supplier's empathy, clarity, and advance planning—provides a competitive advantage over other suppliers who have not developed the same working relationship. It engages the distributor in real discussions about what can and should be done. It provides a forum that permits heated disagreement, then compromise and solution. It avoids the "false courtesy" that often masks outright refusal to agree.

Communicating with distributors is difficult because of the extremely large volume of communications that they receive. The supplier's objective must be to determine how to get a message through that will be received and acted upon. This generally means communicating information concisely and infrequently to the persons who will make decisions based upon it. Suppliers should communicate with these people only when necessary, keeping communications simple and easy to deal with and being very succinct. Distributors will appreciate the courtesy and brevity, and this will pay off in the long run.

JOINT CALLS

A "work-with" is a visit by a supplier that is spent working with one or more of the distributor's marketing or sales staff members. Most often, this time is devoted to making sales calls together on clients. From the supplier's perspective, the purpose is to obtain information on market conditions and trends that will aid the supplier in adapting its market strategies, in collaboration with distributor management. The opinions of the distributor representative are important in developing a comprehensive picture of the market, but they need to be supplemented by the assessments of customers and others in the trade. Work-withs are also used to train sales personnel on the selling points of the supplier's products.

Suppliers appreciate work-withs because they learn what is going on in the market. There are no reliable forms of intelligence that substitute for such lessons. The large market survey firms provide complementary data and are getting more detailed information, but their information cannot yet compare with the rich information gained from being on the street with sales representatives, retailers, and customers. The winery representative will learn a great deal by listening to what others say about the business or the market.

Work-withs are scheduled in conjunction with the distributor and are normally in accord with a preagreed arrangement. Other visits may be to discuss results with management, to arrange special events, or to plan new sales initiatives. Whatever the purpose, visits and work-withs are invaluable as business builders and learning experiences, but they should be scheduled only when a specific objective can be accomplished and when both the supplier and the distributor can afford the time and cost required.

Ground rules for joint sales calls should be agreed to before they are undertaken. Some rules to be included are the purpose of the visits, who is to be visited, what products are to be featured, what is to be communicated, and what is to be learned. Two of the most important behavioral rules are not to discuss pricing with distributor representatives or customers, and never to preempt the distributor representative. Winery representatives should avoid the temptation to be a "big shot" or the "problem solver" by offering a pricing solution. The winery representative may have the actual or perceived ability to change pricing, through discounts and allowances, for example, and will be tempted to do so in the hope of obtaining added business. However, such decisions preempt the distributor's authority over customer prices as well as management's prerogatives concerning supplier prices. Such preempting sets a bad precedent and may be illegal in some markets.

Other guidelines and rules will help as well: the winery representative is a guest on the call; the winery representative should never usurp the role of the distributor representative or criticize the person in front of customers; the winery representative should not try to fix things unless specifically asked by the distributor representative; the customers belong to the distributor, not the winery; and the distributor representative should not be selling other wines when on a joint call.

It is a good idea for the winery representative to summarize calls and discuss them with distributor senior management before ending the visit. This provides the opportunity to discuss possible alterations to sales strategies or messages. Any evaluations and recommendations (for example, in where the product is positioned on retail shelves) should be worked out with the sales representative before bringing them up with management.

EVALUATIONS

The winery has the right and obligation to evaluate distributor performance and measure how well mutually determined goals have been achieved. The evaluation process should be objective and open, and the results should be discussed with the distributor. The objectives are to focus on problems rather than people and to identify alternative ways to resolve the problems. Evaluations that only describe problems are far less useful than those that suggest solutions. Laudatory evaluations may not be very useful either because they are not needed for new decisions. If the evaluation reveals a failure that merits a penalty under the distributor agreement, the winery should be sure that the penalty is applied equally to all distributors with similar failures. Discrimination or inconsistency will be a very expensive litigation point if a "wronged" distributor can prove that another distributor was treated less harshly for a similar failure.

A business review should be scheduled annually as a complement to periodic evaluation sessions that might occur after joint visits. The effectiveness of such a review is largely determined by the quality of the prework done by the winery in support of its performance assessments. Evaluations, even in the face of failure, are infinitely easier if the winery and distributor have spent a lot of time up front to define what success should look like. With such explicit understanding at the beginning, the partners do not get into much heated discussion when determining if objectives have been met. It is amazing how differently people can perceive a situation if they do not have an explicit understanding at the start.

These reviews do not have to be more involved or formal than both partners feel necessary, given the state of business in the given market. In other words, a winery should not waste the distributor's time if business is good and the winery is satisfied; however, a focused, formal review is needed if performance is falling well below mutually determined goals.

In setting up work-withs, informal evaluations, and formal reviews, the winery should focus on markets where existing or potential business is most substantial and where performance is weakest. This simple point is often missed because it is difficult to avoid the tendency to visit "friendly" markets and distributor "friends" and to evaluate performance based on such visits. This is often revealed in sales reports that show inadequate coverage of problem markets.

TERMINATION

Termination is a last resort in distributor relationships. It is expensive and difficult, but it is justified where failure is clearly due to distributor shortcomings rather than the product, winery activities, or market failures. An experience in Europe with a spirits distributor provides an example of how much a termination procedure can cost. I had a case where a distributor in a very small market in Europe had not achieved any of the agreed-upon objectives over a period of years. It was verging on bankruptcy and well known to the trade as an underperforming distributor. I counseled the distributor and carried it for a year and then had to terminate it. The payoff to the distributor was $500,000 to avoid going to court in the distributor's country. This was not for a large brand, but it suggests the liability that can be incurred in termination. Terminations inevitably lead to compensation being paid to the terminated distributor.

The distributor agreement should have a set term and specify how the agreement can be terminated at the end of the term. Such termination is easier than one undertaken while the agreement is still in force. For this reason, the agreement should describe reasons for termination during the term of the agreement. Legally, these can include failure by the distributor to meet agreed-upon goals or to conduct agreed-upon activities; or discriminatory pricing, sales representation, or service arrangements that harm the winery. However, in practice, defense against termination may cite failures of the product or of the market, or activities of the winery. In some cases, termination is clearly the best alternative, but in other cases there may be preferable options.

The winery should analyze the factors that led to a termination, rather than just building a case based on distributor failure without investigating the cause. The information gained can be significant in evaluating other distributors or in setting up a new distributor arrangement. It will help the winery to avoid repeating past mistakes.

Some wineries believe that problems of termination and other actions covered in the distributor agreement can be worked out through distributor councils. Such councils are organized by wineries and include all or most of the distributors with which the winery works. The objective is to exchange views on common problems in winery-distributor relationships and to explore better ways for marketing the product. These councils can become winery-financed social functions and, at worst, a forum for collusion among competitors. The choice as to whether to institute such a council should be made after a careful review of its costs and what the winery can realistically gain from such an activity.

CONCLUSIONS

The problems that really drive distributors crazy are things that the winery can largely control. The operation must be streamlined, and the winery must be shipping on time; the winery must not be running out of stock; pricing must be consistent and easily communicated; and no one should be usurping the distributor. These day-to-day operational issues are more important than often-cited concerns about lack of advertising support or excessive pressure to sell. If the fundamental problems are resolved, then it is far easier to deal with subsequent problems.

The primary problems are generally overcome by being very direct, up front, and businesslike. Distributors want a succinct, well-thought-out expression of how the relationship should work in their market. That takes a lot of prework on the winery's part as well as clarity in communications. Without this investment of time and effort, the distributor and winery may be in court before the wine is in the store.

The ultimate advice to winery managers is to understand the business, the markets, and the distributor's needs and to accept the idea that if the distributor does not make money, the winery is unlikely to either.

CHAPTER **25**

A Changing Distribution System

Ed Everett

INTRODUCTION

The U.S. wine distribution system has changed substantially since the end of prohibition in the United States in 1933, and it is still evolving. These changes significantly affect the marketing of wine. Wineries, distributors, and retailers use this system; depending on how well they use it, they will either lose or make money. Consequently, it is important to know what changes have occurred in the wine distribution system and how they have affected marketing decisions. This chapter provides a review of the more important changes and their implications for users of the distribution system.

THE EARLY PERIOD

In the congressional debate about the repeal of Prohibition, nobody could agree on anything. Everybody knew that gangsters had taken over the alcoholic beverage business, Americans (most of them, at least) were breaking the law by drinking, and Prohibition was a failed experiment that had to end, but many states wanted to stay "dry." Georgia was so determined to keep the "devil's brew" from its sacred soil that it threatened to prohibit the making, selling, and transporting of alcohol in the state. This

---

Ed Everett enjoyed a long career as a California distributor, importer, and industry observer and was involved in building many major brands during his career. He passed away in 1999, but his company, New World Wines, is still run by his wife, Anna, and his partner, John Anderson.

flew in the face of the Commerce Clause of the U.S. Constitution, which prohibited individual states from impeding interstate commerce. In the face of this opposition, the U.S. Congress compromised in order to pass the Repeal Amendment to the Constitution. It agreed that the Repeal should include a provision allowing individual states to decide whether and how alcoholic beverages could be imported and distributed in their territories. This became the basis for dry states and dry communities within some states. In return for this authority, the opposing states agreed that the Commerce Clause should apply to shipments of alcoholic beverages through any state.

Once given the right to determine whether to maintain or repeal Prohibition, most states decided to keep Prohibition. Almost all the southern, midwestern, and mountain states, and some of New England, stayed dry. California, Illinois, New Jersey, New York, Massachusetts, Nevada, Oregon, Washington, and Wisconsin went wet. This resulted in a checkered distribution system and a nightmare of different regulatory systems. It was not until the end of the 1930s that most of the states dropped outright Prohibition, but they tended to retain state distribution systems or to otherwise restrict distribution of alcoholic beverages. States like Utah enforced a complete state monopoly, except for some semiprivate licensees in the restaurant business and in ski resort areas. Pennsylvania still maintains a monopoly on distribution through state-owned warehouses and retail stores. New York does not allow the sale of food and wine in the same retail premises. Other states protect their distributors by prohibiting wineries from selling directly to customers and requiring that wineries work through distributors licensed in the state. These rules reflect the influence of pressure groups, the power of existing public and private organizations, and the flow of money generated by taxes and distribution profits. Alcohol is big business, and it begets big power.

THE MIDDLE PERIOD

World War II produced very stringent regulations on the production of alcohol. Wineries and distilleries had to produce alcohol for the war effort. There was little industrial or distribution development under wartime rationing and product diversion. After the war, a huge burst of economic activity took place and new distribution structures began to develop. The big distilleries came out of the war with bulging treasuries and began to set up distribution houses that became money machines. The houses were huge enterprises, most of which still exist. They lobbied for laws that protected them and made it very tough to enter the business. The kinds of wines popular today played almost no role in this era. The wines were mostly high-alcohol dessert wines, one of the cheapest sources of alcohol. That was where the volume and money were and where the large liquor-dominated distributors wanted to be.

A more wine-friendly distribution system began to emerge over time as the Gallo Winery of California built a network of distributors. The winery disliked the liquor houses and their attitudes toward wine. It started working tentatively with beer distributors and found that they could be trained to do a better job with wine. Eventually, the winery began to recruit and train its own distributors. Ernest Gallo and the people who worked with him wrote the book on wine distributing. It is full of homilies and little instructions and constant reminders that the business of distributing wine is point-of-sale driven. Eye-level positioning, key visual presentations, use of the cold box, signs, and information all contributed to point-of-sale effectiveness.

The 1960s were transition years. Major producers began developing lighter wines, introducing products like Vin Rosé, Paisano, and Mountain Red. These were wines that aimed to bring consumers over to table wine and away from dessert and sweet wines. Sparkling wines surged on acceptance of Cold Duck, a sweet red sparkling wine. There was the beginning of "pop" wines made from pears and apples. This was the "hippie" era, when preferences went from beef and bourbon to fruit and wine. Wine conjured images of old-style flowers and people reading poetry. People spread the word that wine was good, that it was natural, and that it fit in with the new culture.

The distribution system slowly began to shift. During the 1970s, the distribution houses that Gallo had trained began to dominate the business. They ran rings around the liquor-based distributors, which did not train their people to help with wine displays, to carry feather dusters for cleaning bottles, or to hang up persuasive little signs about what wines went best with what foods. These activities helped sell wine, and they brought Gallo large increases in market share. At the beginning of this period, there were only a few leading brands. Imported wines like Blue Nun, Mateus, and Lancer's were enormously successful and were distributed largely by liquor houses in much the same way that spirits were distributed. These imported wines lost favor, however, as consumer tastes shifted and old-line distributors weakened.

It was easy for many producers to make money during this period partly because of fair trade laws. The laws were put in effect in California in the 1930s as the result of business pressure to ease the impact of economic depression. They subsequently were enacted in a large number of other states. In the case of wine, the laws mandated that wineries must regularly post the retail price and margins for their products. No person could sell the wines for less than the posted price. With protected margins, competition focused on service rather than price, which was set to protect the least efficient distributor. Consequently, there were lots of distributors and retailers, and most of them made money. Consumers, though, lost the advantages of price competition.

THE LATER PERIOD

Big changes came suddenly at the end of the 1970s, setting in motion the changes occurring today. These big changes were stimulated by an interest in deregulation. In 1979 a retailer in San Francisco challenged the fair trade laws in court, causing price posting at the retail level in California to fall apart. The following year, the Mid-Cal Aluminum Company appealed to the U.S. Supreme Court against an adverse ruling under the California fair trade law. The Supreme Court overturned the California law, setting in motion significant deregulation of distribution throughout the United States. Price competition returned, which forced a massive reduction in the number of alcoholic beverage distributors. This was the beginning of the new age of fully competitive and innovative marketing.

Similar reductions were occurring in the retail industry. After the demise of fair trade in New York, the number of retail stores began to drop. By 1985, the number of licensees had declined by one-half. In California, the number of retail outlets also declined, and chains like Liquor Barn came in. Then warehouse clubs like Price Club and Costco (now merged) began to acquire a share in wine marketing. These two merged clubs became the single largest retailer of wine in the United States, wiping out many of the lazy retailers that had been protected by fair trade. Selling became a much tougher proposition for distributors, and some distribution businesses began to fail.

The practices of the warehouse clubs have changed how distributors operate. For example, Gallo's whole distribution concept was based on control of point of sale. It is ironic that the biggest wine retailer in the United States today forbids distributors from doing anything: Distributors cannot go into the store, put up a sign, and or tell the retailer what to do. There is no cold box. The outlet employee takes a pallet of wine from the warehouse and plunks it on the floor—that is the merchandising. The warehouse clubs' approach was a radical shift away from the approach to distribution that focused on service to the retailer.

Another radical shift occurred in sales to restaurants and hotels. At one time, distributors battled for control of the wine list. Typically, a distributor would print and supply the restaurant with wine lists in return for some influence on what wines were listed. There were opportunities to supply the house wine (typically a generic wine) and the other dinner wines. While the distributor had to show some semblance of equity in choosing wines on the list, there was still plenty of opportunity to load the list with the distributor's wines.

Today, the generic house-wine business has virtually disappeared as restaurants serve chardonnay and cabernet by the glass and include a house selection of the same class. Wine lists have returned to the restaurant's control because word processors make the job easy. Now every restaurant worth its salt does its own list. The two biggest forms of control that distributors once had, merchandising for large retailers and wine list preparation for restaurants, have been substantially weakened.

However, as the number of distributors has dropped, the surviving distributors have gained economic power over some accounts through credit terms or merchandise availability. The tradeoff might be a position on the wine list in return for extended credit, or an eye-level position in a retail store in return for guaranteed product availability. The current distribution environment is a far cry from the days of fair trade laws.

Wine distribution in the United States is still complicated by the Byzantine rules governing distribution in individual states. Up through 2000, industry attention was focused on state laws that made it a punishable felony to sell directly to a consumer or retailer in the state without going through a licensed distributor in the state. These penalties are a barrier to Internet sales of wine in several, if not many, states. This trend could precipitate a crisis in law similar to the crisis concerning fair trade policies, and distribution strategies will need to adjust to whatever changes emerge.

Distributors also gained power over wineries as the number of wineries seeking market access grew rapidly. The emergence of these wineries, mostly competing for distributors, retailers, and consumers, led to intense efforts to differentiate producers and products from one another. This resulted in a large increase in the number of brands and the rate at which they were introduced into the market. Well-established firms began introducing new brands. These strategies are analyzed in detail elsewhere in this book. Their impact, however, was to give a competitive advantage to wineries that could create new images that raised the interest of consumers and the trade. These wineries had an easier time developing profitable working relationships with distributors. The differentiation strategies also led to the growth of direct selling through wine clubs, the Internet, or tasting rooms. These strategies were designed to bypass the "traditional" channels of distribution.

CONCLUSIONS

Wine distribution today still reflects many of the changes that took place in the 20th century, including the enactment of a Prohibition policy that left individual states in control of distribution and consumption. Distribution strategies obviously have to accommodate these diverse rules. For a long time, distributors and retailers were protected by fair trade rules that virtually eliminated price competition and left room in the system for a large number of small, and often inefficient, distributors and retailers. With the demise of government price regulation, large-scale distributors and retailers, with varying concepts about selling wine, came to dominate distribution. Over the same period a large number of new wineries emerged and began searching for distribution. As a result of this change in the distribution structure, the new wineries had to develop marketing strategies that would differentiate a winery sufficiently to attract attention from both consumers and the trade. It is within this very competitive environment that the marketing concepts discussed in this book must be applied.

CHAPTER 26

Distribution Strategies and Legal Barriers

Michael B. Newman

INTRODUCTION

This chapter examines distribution strategies applicable to the U.S. market from the perspective of international marketers. The principles involved will apply to other markets where regulatory barriers are significant. They are particularly appropriate for marketing policies aimed at building a business in a large set of different countries. The chapter focuses on decisions made by marketing managers about how to distribute product in the United States.

Wine distribution in the United States is affected by laws and regulations enacted and enforced by individual states. Some are complicated and stringent, and some are easy to comply with. Of course, a company does not need to target the entire market. According to MKF WineStats 1998 U.S. Wine Market (published by Motto Kryla & Fisher LLP), approximately 60 percent of the wine sold in the United States is sold in 10 states, of which 5 account for 50 percent, and 1, California, accounts for about 20 percent. A foreign company can enter the U.S. market on a fairly limited basis but still reach the majority of the market.

The chapter examines various strategies, including using an independent U.S. sales agent, importer, or national distributor; establishing a U.S. marketing company; or establishing a direct sales or import operation. The

Michael B. Newman is the resident partner of the San Francisco office of the law firm Buchman & O'Brien, which specializes in alcoholic beverage law. His practice includes counseling industry clients on licensing, distributor relations, and export regulations.

discussion considers handling licensing, choosing a state in which to locate the business, getting the product sold, buying out existing firms, creating a U.S. brand, and initiating a joint venture.

USING SOMEONE ELSE FOR MARKETING

The simplest way for a firm to avoid the legal and marketing risks of doing business in a foreign country is to appoint someone in the country to handle marketing and distribution. In this situation, the winery (or exporter) sells its wine to some U.S. company that acts as the winery's agent or representative. The U.S. company is responsible for importing and marketing the wine. It will hold all the federal and state licenses and permits, and it will handle all the registrations, price postings, and other requirements that must be met to sell wine. In a few instances when the winery is shipping into a particular state, it will need a license in its own name, but the U.S. company should advise on that requirement. Basically, it is up to the agent to deal with these different regulatory barriers. This distribution option will probably help the winery avoid most potential U.S. tax issues, but it will limit the amount of involvement and control that the winery has over marketing strategies. The winery should provide advertising and promotion and focus on differentiating the product and package in order to back up the agent's program. But in the final analysis, success is dependent primarily on the U.S. agent.

The winery may hire an employee or independent contractor to be its representative to monitor the agent and the market. The representative's primary responsibility is to oversee what the agent is doing and evaluate performance against competitors. Another role is to convince consumers to drink the winery's product. The process is to work with the agent and to be in the marketplace, creating an interest in the product among distributors and retailers. In summary, this distribution strategy includes the appointment of a U.S. company to take care of all the selling and the marketing, and the appointment of a representative to oversee that company.

A variation of this strategy is to have several different regional companies as agents. This provides expertise in different markets, which may result in more appropriate marketing strategies. It also may provide more intensive coverage in important regional markets than a single national company could provide. This alternative is more difficult to manage, and for that reason it is probably more expensive. It is probably easier to go the route of having one single agent, but sometimes it is hard to find a good company that will spend sufficient time on the brand.

Another alternative to having a sales agent or U.S. importer/distributor is to have a broker. Legally and practically speaking, there is a big difference between brokers and sales agents, importers, and distributors. The broker does not take title to the product and is paid on commission. The winery still sells to various distributors throughout the country, as arranged by the broker, and the broker receives a commission on the sales. Usually the commissions range from 10 percent to 20 percent, and customarily they center around 15 percent to 16 percent. The advantage of using a broker is that the winery will have more control over the selection of distributors and the price that it pays. The price will be that of the winery rather than that of the importer. The disadvantage of having a broker is that the winery will have to deal with more licensing and brand registration requirements than if working through a national importer or distributor.

Another option is for the winery to negotiate or deal directly with retailers, perhaps giving them exclusive representation in their market. This will require using a "clearing" wholesaler because virtually all of the states prohibit direct "sales." Many retailers do business in this manner. Even if a retailer comes to the winery to order wines for its stores, the winery may have to sell through a wholesaler to get the wine to the retailer. This process is called clearing.

Typically, a wholesaler is supposed to be actively involved in the promotion and merchandising of a winery's brand within the wholesaler's market. This includes storage, delivery, tastings, displays, advertising, and invoicing. However, a clearing wholesaler does very little except invoice and collect money. Perhaps it may physically deliver the wine, but in some cases, a carrier can bring it right to the dock of the retailer and bypass the wholesaler. Commissions range from $1.50 to $6.00 per case just for clearing.

In some markets, such as California, retailers can hold both a retail license and a wholesale license. This permits retailers to bypass wholesalers in acquiring products, and wineries to solicit and sell directly to California retailers. It is largely wine specialty outlets that use this procedure. Although the procedure is used more frequently than it once was, it is still not the main way business is done. In most of the country, wine must be sold to retailers through wholesalers.

Direct selling to consumers is another method of distribution. This is a highly specialized strategy and is often used in combination with other strategies. It is realized through mail order, telemarketing, wine clubs, and the Internet. Use is fairly restricted because of the many state laws that prevent direct sales to consumers. However, where allowed, direct sales to consumers have generally worked well. Selling through the Internet is a recent innovation that appears promising. A wine mail-order business may obtain a retail license in various states so it can sell wine directly to consumers. Some companies, such as Geerlings and Wade, have been quite successful with mail order. States are becoming more liberal in allowing these alternative methods, as long as taxes are paid and regulations are followed.

ESTABLISHING A U.S. MARKETING COMPANY

A second method for getting wine into the United States is to establish a subsidiary U.S. marketing company to do all the marketing, promotion, and public relations and to deal directly with the winery's importer or agent. The U.S. sales agent or importer will continue to import and distribute the winery's product. This is a dual system, with the subsidiary doing the marketing and the importer or distributor doing the sales. The advantage is that the winery does not have to set up a sales force and distribution network in the United States. It is able to monitor the market, supervise the other agents, and develop marketing strategies to benefit the brand. At least one person, presumably the brand manager, interacts with people in the import and distribution company and may interact with distributors and retailers. The principal focus of the subsidiary is on creative activities such as marketing, advertising, and public relations and dealing with agencies helping with these efforts. One advantage is that the marketing company does not need the licenses required of sales and distribution companies. The selling, shipping, and invoicing—not the marketing, promotions, or public relations—create the license requirements. The marketing company will also avoid federal and state taxes that would otherwise arise if it were selling product in the United States. The key distinction is that the importer/distributor is selling the product, while the marketing company is only promoting it.

ESTABLISHING A DIRECT U.S. SALES OPERATION

The most complex alternative for gaining a presence in the U.S. market (or other foreign markets) is to establish a subsidiary company to handle importation, distribution, sales, and marketing. Such a strategy involves the winery fully in the distribution chain. This alternative is the antithesis of working solely with an importer/distributor. It provides maximum control over pricing, distribution, and marketing. The winery itself will have to decide whether the extra control will lead to marginal revenues that exceed the marginal costs of running such a system.

The usual procedure for establishing a direct operation in the United States is to organize and incorporate a U.S. subsidiary company. For various tax and liability reasons, establishing a subsidiary company is generally better than establishing a branch of the company. A company may incorporate in any state, and the fees for doing so vary from state to state. It may then do business in other states, subject to registration fees and taxes imposed by those states. For example, if the company wants to do business in California, the company must be "qualified" according to state regulations, but it is not required to incorporate in California. The company would also have to "qualify" to do business in New York State and pay similar fees. Depending on individual circumstances, it is probably just as well to incorporate in the state where the company does most of its business. Tax structures affect location decisions, and these need to be balanced against the marketing advantages of alternative locations. Minimizing taxes may lead to minimizing sales as well. Consequently, the choices of incorporation and location need to be evaluated from both the marketing and tax liability perspectives.

The company will need a federal importer's permit, a license in the state in which it is doing business, and licenses in nearly all states into which the company ships wine. The latter licenses are not as extensive as are the federal permit and home-state licenses. The company may need to register its brands in addition to obtaining licenses. In many states, the company must file monthly reports of shipments and may be required to post prices regularly. Most companies that directly distribute their products nationally have one or more full-time people to make all the required filings.

The company must decide where to locate its business. As noted previously, this does not need to be the state in which the company is incorporated. The decision depends on tax structures, the prestige of the location, residency requirements, licensing costs, access to market, and ease of travel.

The taxes of concern include corporate and individual income taxes, inventory and other property taxes, sales taxes, and various incidental taxes. These taxes differ from state to state; California and New York tax rather heavily, while Nevada does not tax heavily. The prestige of a major metropolitan area can influence consumer acceptance of a product. In the past, it seemed most foreign-owned companies set themselves up in New York. That was considered a prestigious place to be for an importer; every wine label had to indicate the name and address of the importer. In addition to New York, prestigious locations now include San Francisco, Los Angeles, and Chicago. A label that reads "Imported by Viceroy Imports, San Francisco" is likely to be viewed more favorably than a label that reads "Imported by Viceroy Imports, Butte, Montana." Residency requirements for licensing and other legal matters may affect the location decision. These are the sorts of requirements that may not be immediately obvious but can be discovered through a diligent search prior to the location decision.

License fees influence where the company does business. In some states, it is extremely expensive to get an importer license. In New York State, it is about $7,000; in New Jersey and Rhode Island, it is roughly one-half of that; and in California, it is only a few hundred dollars. Market access is another important factor. It is reasonable to be in a state with a major wine market, such as California, New York, Illinois, or Texas. Working in these high-volume locations reduces unit selling expenses because transport costs and shipping times are lower, the sales force can be more productive, and market monitoring is easier. Another consideration may be the ease of travel between the foreign parent company and the U.S. sales subsidiary. For this reason, Australian companies often set up their direct operations on the West Coast and European companies set up their direct operations on the East Coast.

The company must decide how it will organize to sell its products. This decision depends partly on the brand strategy chosen by the company. For example, there may be a strong argument for creating a U.S. brand to differentiate the product. Perhaps the home-country brand name may be difficult for Americans to understand or may suggest the wrong image in the United States. When starting with a new U.S. brand, the company should move slowly before establishing a large and expensive organization.

There are two prominent choices. One is to organize a national sales force and undertake the job of a national distributor, and the other is to utilize regional brokers for distribution. Assuming that the company has the volume to justify significant overhead costs, it can hire sales representatives or managers in different markets to promote the product. It costs an estimated $125,000 in salary, benefits, overhead, and travel to employ a single sales representative. Thus, the cost of having a national sales force can easily reach over $1 million dollars annually, requiring revenues well in excess of $20 million to justify it. For a wine selling to distributors at $100 per case, this implies movement in excess of 200,000 cases.

A far less expensive way to do business is to have regional brokers in principal markets to represent the company. Brokers are not exclusive representatives and handle other brands in the same market. They usually work on commission and do what sales representatives do, although brokers also devote time to other brands. It is possible to find brokers with noncompeting brands that need coverage in the market niche targeted by the company. The advantage of this arrangement is its lower cost and flexibility. Flexibility is gained because it is easier to change a broker than it is to fire an employee. There are no laws that protect brokers in the way that employees are protected. Most broker contracts include a 30-day termination notice unless a different arrangement is agreed to at the outset. One disadvantage is reduced control over the sales force and over sales decisions. Another is that the company's brand may not be sufficiently important to the broker to motivate extra sales effort or attention. The lack of exclusivity is one of the key problems with having a broker network. To deal with this, some companies combine a sales force with a broker network. They may have employees in the state of California and brokers in other markets, or they may have a key salesperson in New York, Illinois, and the Northwest; and brokers in the Southeast, the Midwest, and the Southwest.

Foreign companies also have the option of buying out an existing importer and national distributor. This provides instant coverage with experienced personnel. This strategy runs the risk of incompatibility in management styles, which may be detrimental to sales.

Joint ventures with other distribution firms is also an alternative. They require careful planning in order to identify and agree upon common objectives. But there may be problems with competing management styles and questions about which partner is "really" in control.

CONCLUSIONS

Three important conclusions can be derived from this chapter. The first concerns a strategy for a high-volume company. If a company wants to build a national brand and develop sales of 750,000 to 1 million cases throughout the entire country, it should either set up its own sales and import company or use a national importer/distributor. In the latter case, it should maintain a presence through a marketing company or some full-time marketing representation. At minimum, the company should have one-half of the control over the business. In order to move high volumes, the company should have control over the marketing function at least, and the sales function also, if feasible. There is less risk in leaving sales to experienced local personnel, but marketing is a company responsibility that is based on principles that apply worldwide. Increasingly, there are greater similarities between marketing in the United States and marketing in Europe or South America. But there are still large differences in selling practices.

The second point concerns focused regional marketing and sales efforts. Most of the wine sold in the United States is sold in relatively few markets. If a company aims to build a major brand it can concentrate on specific regions rather than attempt national distribution. In this instance, the company can build the business through regional importers or regional sales agents and avoid having a single national distributor.

Finally, a small boutique supplier with a high margin and limited quantity of high-quality wine will find that a very limited U.S. presence can be profitable. The best strategy is to work with a limited number of retailers in various markets that will arrange for a wholesaler/importer to clear the product for the supplier. The direct marketing business has unlimited potential and will undoubtedly expand over the next 10 or 20 years as demand increases and laws are amended to ease the change.

CHAPTER 27

Selling Wine in and to Supermarkets

Robert D. Reynolds

INTRODUCTION

Supermarkets are a major outlet for wine. They, and other grocery stores, account for an estimated one-third of U.S. wine sales to consumers. They are the most important channel in most of the large wine-buying states and provide constant consumer exposure to wine. They feature easy access, are more comfortable for women than many alternative outlets, and are innovative marketers. Consequently, it is important for wine marketers to learn how to sell to supermarkets effectively. To do this, they must understand what motivates supermarkets in selling wine. This knowledge is essential for persons selling wine in supermarkets, or for those selling it in other outlets serving or competing with supermarkets.

The objectives of this chapter are to analyze how wine is managed in supermarket outlets and to discuss how this affects the work of those selling wine in and to supermarkets.

WHAT SUPERMARKETS OFFER

Wine is sold in supermarkets in almost every highly populated U.S. state, with the exception of New York, Connecticut, and New Jersey. Sales

---

Robert D. Reynolds is a marketing economics consultant specializing in consumer products and the grocery business. During his 33-year career, he has worked for major food marketers, retailers, government agencies, and marketing trade associations. He formed Reynolds Economics in 1987. In addition to marketing assignments, Reynolds Economics provides support for issues in litigation.

are restricted in Colorado and a few other less populous states. Supermarkets offer constant consumer exposure. People go to the grocery store more than 2.5 times per week on average, and this provides opportunities to see a lot of wine. This exposure is facilitated by the long hours of operation by most supermarkets. Many are open for 24 hours, and although they may not be able to sell alcoholic beverages during certain hours, they still provide exposure. Women are the predominant shoppers in supermarkets and the principal purchasers of wine there. Thus, the supermarket becomes an essential outlet in serving the female segment of the wine market. Supermarkets are also important in establishing new merchandising techniques that later show up in other outlets. This provides an opportunity for wineries selling to supermarkets to anticipate trends that may show up elsewhere.

MANAGING THE SUPERMARKETS

Management in a supermarket is complex. Therefore, it helps to know what motivates management decisions. One of the factors is the impact that a product decision will have on incremental sales. Supermarkets are always looking for ways to make the most use of the available shelf space in the store. One test for the allocation of shelf space is the impact on incremental sales of using one product or brand rather than another. If it is simply a tradeoff, the chance to get a new product into the store is slim. The challenge for the wine marketer is to convince the supermarket that the incremental sales and profits from the wine will be better than those from existing products. The wine marketer must think in terms of what product will be displaced if the new product is taken on, or if the space allocation for an existing product is expanded. This mindset will lead to calculating the differences in sales and profits that will occur if the change is made.

To boost incremental sales, retailers look for presold or well-marketed products with an associated advertising and promotion program. This may include cooperative advertising, a coupon program, labels that pop out of the displays so they can be read, and sales service. Perhaps the sales representative should be in the store dusting off the display so the supermarket employees do not have to do it. The marketing program associated with the product should put advertising, promotion, labeling, and packaging together in such a way that they are elevated to a position equal to that of the product itself. It is essential that the program associated with the product be compatible with the image of the product. Otherwise it will undersell or oversell the product. In my opinion, the quality, taste, and functionality of the product account for about 25 percent of supermarket managers' decisions to place wines in stores. The nature of the package is the next most important attribute, and then everything else falls in line. But a very good product simply will not do the job alone.

Sales support services are something else that retailers want. If the competitive product offers shelf space management service, it is more valuable to the supermarket than a product that is not bundled with services. If it is important for the winery's product to be well presented to the consumer, then the winery had better arrange for that presentation. Otherwise the product will not be presented as well as a competitor's product and will not sell as well. A dusty bottle of wine, although it may have great appeal sitting in the cellar or in caves, will be a disaster in the supermarket.

Supermarkets decide to stock new products because they offer the prospect of increased profits. Profitability may be measured in percentage terms as the difference between cost and selling price. A measure of more importance to the supermarket, however, is the return in dollars per unit of shelf space per month. This measures the return that a supermarket earns by renting shelf space to various products. It combines gross margin with the velocity with which the wine moves through the store. The wine marketer has to demonstrate that the combination between gross margin and velocity will return more shelf space rent than that of the product being displaced.

Shelf space decisions must recognize the necessity for a supermarket to offer a selection of products. Salt selling for 15 cents per pound does not return much profitability, but the store must have it in order to attract customers. There are other necessities that probably do not pay their own way but cannot be dropped. In consequence, the shelf space rental for nonessentials must be higher than that for essentials. Decisions about what to stock are also influenced by the costs of keeping a product in the store. An extensive display of wine that is managed and maintained by sales representatives is less costly to the supermarket than, say, a private-label frozen product that store employees must stock and maintain.

Supermarkets, by and large, are not experts in marketing wine, so they are looking for a marketing partner. This is not surprising, because retailers stock 15,000 to 20,000 items in a typical store. Retailers like it when sellers tell them how to sell products effectively and help the retailers do it. Stores might partner with a distributor, or a winery, or a major brand. The function of the marketing partner is to provide marketing expertise and honest advice on category management, not just a sales spiel. The winery or distributor needs to provide product support throughout, until the product is bought by the consumer. Success in marketing to the grocery business comes from not just selling the product to the retailer but supporting the retailer. This includes getting the product appropriately placed on the shelf, ensuring that displays are well maintained, attracting the attention of the retail consumer, seeing that the product is actually sold, and ensuring that the consumer is sufficiently pleased to come back for more.

The elements of a product support package can be classified according to their use "backstage" and their use in store. Backstage services include informed product presentation, category management, and market information. Service and support in the store include shelf space planning, product introductions, new sets and resets, and shelf maintenance.

There are numerous deals that one can offer in making a sale to a retailer, such as slotting allowances, special deals, free goods, special displays, and coupons. But if the retailer is not supported and does not get the turn on the product and repeat sales, the initial sale will have been a waste. The power of information is increasingly important, and sellers can often use it to guide retailer decisions. One source is grocery store scanner data compiled by IRI or Nielsen Corporation. Such information is absolutely critical in selling to supermarkets. For example, knowing that scanner data show dropping sales in a particular category, a winery with products in that category can work out a sales appeal based on the winery's high reputation in that category, encouraging stores to keep its product and drop other products instead. In other situations, it can prepare a sales pitch that capitalizes on its foresight in having a preferred brand in a growing category.

Wineries and distributors need to own the marketing process and take responsibility for it from beginning to end. They need to deal with problems such as corks being difficult to remove. If the winery does not take responsibility and solve the problem, it will lose sales. The consumer is unlikely to buy a fancier cork pull and will probably flip open a can of beer instead. Supermarkets also expect wineries and distributors to provide category expertise. This, of course, opens the opportunity for category management, as discussed in Chapter 16. Retailers expect a program sell rather than just a product sell. A program is built around advertising, marketing, coupons, shelf talkers, and care for the items in the store. Expected services include dusting the bottles, arranging to accept returns after the holiday season, and helping with displays.

All supermarkets do not have the same needs for their wine sections. For example, one chain may deal predominately in regions where wine preferences are toward sweeter wines in large bottles, while other chains may be dealing only with varietal wines, predominately in 750 ml bottles. These differences will also occur within different stores of the same chain. Persons responsible for wine purchases in some supermarkets may have different degrees of experience, knowledge, and authority than the wine buyers working for other supermarkets. These situations call for sales approaches that are tailored to the needs of the specific chain or group of supermarkets within the chain. The lesson is that a single sales strategy for the supermarket sector is likely to be ineffective.

MANAGEMENT THEMES

The major theme in grocery management is efficient consumer response (ECR). ECR is concerned with finding ways to serve the consumer more efficiently—in other words, getting good products where they need to be, when they need to be there, and at the right prices. ECR is the antithesis of slashing and burning policies. It concentrates on products and services that provide value to the consumer and eliminates those that are superfluous to the consumer. Its objective is to obtain an optimal inventory that will produce sales and profits. The retail process is being substantially reengineered to follow the principles of ECR.

The current management style is far different from how it was just a decade or so ago. There used to be very intuitive managers who were in charge of most of the marketing process. They worked by touch and feel, largely because they did not have the information that is available now through scanning processes and consumer audits. Now the people responsible for marketing are much more likely to be very analytical, with a great deal of information at their command. The information is derived from proprietary store records and from the common data sources such as Nielsen and IRI. These data-driven systems support sophisticated management tools that literally map the retail shelf for maximum productivity.

These systems apply complex software to a huge amount of data to come up with a set of rules or a set of schematics to guide decisions about product selection, pricing, and placement. These are decision support tools. They do not make the decision; they provide the information that will help the manager make a better decision.

The other new trend is in transaction and productivity systems. These are systems—such as electronic data interchange (EDI)—in which computers in each store talk to computers at headquarters, tracking the sales of each selected item. When inventory reaches a predetermined level, the computer generates a purchase order that reorders automatically from the distributor, the winery, or other vendor.

An important function of the ECR system is the elimination of nonproductive inventories. This is not inventory reduction but an attempt to build inventories of desirable products that provide a sense of abundance in a store. The focus is on products that move—at the expense of those that do not. The philosophy of the 1980s in major supermarkets was to have a huge variety of wines, even if a lot gathered dust on the shelves. The cabernet 43rd on the sales ranking was taking up as much space as the cabernet ranked number 11. The low-selling cabernet was a logical candidate for elimination under ECR.

A major goal of reengineering is to reduce costs. It involves changing the way business is done and establishing systems that help guide decisions toward optimal results. One of these is a management information system that provides needed data and software to analyze them in a meaningful way. The data-based decision support systems include ECR and category management, both of which are discussed above. They also include activity-based costing and continuous replenishment systems.

Data-driven transaction and productivity systems include EDI and "just in time" deliveries. The first is a fairly new system that is expanding rapidly. The latter has been part of management for the better part of 20 years, although it has been tremendously refined as better electronic equipment appears. Another example is vendor-managed inventories, discussed elsewhere in this chapter. This system involves shared sales and inventory data, regular and promotional orders, and vendor inventory responsibility. A system likely to be used more frequently is that of invoicing the retailer for products as they are scanned. This is an electronic version of consignment selling.

CONCLUSIONS

The important factors on the horizon for supermarket management are enhanced vendor-managed inventories, Internet data exchange, computer-based home shopping, performance-based promotions, and an expanded role for information. All of these are present now to a more limited degree. They are expected to become far more significant. The most important of these is enhanced vendor-managed inventories, in which the vendor is actually in charge of stocking the shelves. A notable current example is consignment selling. For example, a major premium ice cream manu-facturer had an arrangement with a large retail chain through which the vendor had total control over its shelf space. The vendor brought the prod-uct into the store, put it on the shelf, took out old stock, and rearranged the display. The chain still determined product pricing. The vendor was also in charge of product that did not sell. If the vendor decided to bring new product into the store, the vendor had to place it within its allowed space and had to take it back if it did not sell, without invoicing the chain at any time. The products are not invoiced to the chain until they are scanned at the checkout counter.

Premium wines are specialty products that could be the subjects of a vendor-managed inventory agreement, assuming that any legal hurdles are set aside, as they might be if the commercial incentives are strong enough. This could result in an agreement in which the distributor or the winery is fully responsible for the productivity of its shelf space, including determi-nation of which products are stocked. This is almost a rent-a-shelf kind of situation. It is the retailer saying "I have real estate that has exposure to consumers, and you can have a piece of it if you promise me profitable rent." This arrangement will gradually become more common.

The use of the Internet to exchange information is surging. It looks like more of the same in the future as users discover the value of home pages and downlinks. The Internet is also fostering home-based shopping, which will affect supermarket strategies. A major grocery outlet exclusively for Internet shopping opened in the San Francisco Bay Area in 1999. There will be a continued improvement in the availability and usefulness of data and information. This will occur partly because of even more efficient computers and analytical programs, and partly from the public and private sectors catching up in the information race.

CHAPTER 28

Selling Wine in and to Large Specialty Stores: The Case of Beverages & more!

Steve Boone

INTRODUCTION

Our research indicates that stores featuring wines and related products account for about 20 percent of total wine retail sales. The stores may range in character from a corner liquor store to a regional wine specialty store. Within this group are some important specialty chains that compete favorably with supermarkets and discount club stores. It is important for wine marketers to know what motivates these firms, how they select suppliers, and how they sell to consumers. This information can then be incorporated into marketing strategies that help suppliers do a better job of selling to retailers and help retail managers do a better job of selling to consumers.

This chapter presents a case study of the California firm Beverages & more! (B&m), discussing the marketing principles that underlie the firm's operations. The purpose is to provide a better understanding of how these principles are applied. The chapter considers why the business was organized, what it looks like currently, how the company markets itself, how its practices differ from those of other retailers, and what the future holds for the firm.

Steve Boone began his wine career as a buyer for Safeway in 1973. In 1979 he founded Liquor Barn, later served as CEO of Cost Plus, and founded Beverages & more! in 1994. *Market Watch Magazine* has named him a retail superstar, and *Wine Spectator* listed him as one of the 100 most influential people in the world in the wine industry.

AN OVERVIEW OF B&m

B&m opened its first store in 1994, and by 1997, it had 20 outlets, 13 of which were in Northern California. Expansion continued through 1999. As a chain operation, B&m is not large by supermarket standards, but with an expected gross of over $200 million, our records show that B&m was the second-largest beverage-only chain in the United States in 1999.

The spacious stores average about 18,000 square feet (1,672 square meters). They offer a fairly wide variety of products and look something like supermarkets. They are clean, well lit, and shoppable. About 40 percent of B&m's customers are female, a much higher percentage than attracted by the average liquor store. The average customer is 45 years old, makes about $105,000 a year, and lives in an urban area. B&m has positioned itself as a high-end retailer offering a broad selection of beverages, related products, and specialty food items at low prices.

Wine accounts for about 40 percent of total revenue. The selection is very good and varies from store to store to appeal to local preferences. The Florida store carries twice as many imported wines as the California store because the market for imported wines is much better in Florida. The San Diego stores offer wines from the nearby Temecula regions, but the San Francisco stores do not. Overall, an average store stocks about 370 chardonnays from California, 310 cabernet sauvignons, and a good selection from Bordeaux, Burgundy, and other wine-producing regions. Some wine shops may have a better selection, but the B&m selection is massive compared to that of most retailers.

Beer accounts for about 15 percent to 16 percent of retail sales. B&m carries about 1,000 beer SKUs (stock-keeping units: individual line items, including varying container sizes of the same brand and type). There are a little over 300 imported beers, more than 300 domestic craft brews, and most of the major brands such as Budweiser, Miller, and Coors. The emphasis is on craft brews and premium imports. Spirits are about 20 percent of the business. The chain carries about 2,000 different items, including 100 single-malt whiskeys; a large array of gins, vodkas, and more exotic beverages; and a great selection of cognacs, Armagnacs, and premium brandies.

Cigars account for about 5 percent of sales. About 300 different ones are in stock and sold from a three-sided walk-around humidor. The humidor is of a unique design, not one of the restrictive walk-in humidors that impede the easy flow of customers.

B&m also sells food and related items through the "& more" part of the business. The offerings are flexible and include a fairly extensive variety of food products: olive oils, pastas, snack items, caviar, condiments, and other choices that relate to the core business. Other products include glassware and upscale wine accessories, from wine racks to $100 wine openers.

Each store has a concierge who greets people when they enter and thanks them when they leave. The concierge does anything a customer wants, from changing a tire in the parking lot to guiding a customer to the best wine in the store. With strategies like having a concierge, B&m has tried to provide an alternative to traditional retailing.

B&m does not carry some of the "brown bag" wines, spirits in pints or half-pints, beer in single cans, and magazines or videos that are not product related. There are no payphones or news racks outside the stores. B&m is trying to discourage customers looking for a quick drink, a brew to ride home with, or a racy magazine. It is also discouraging loitering in and around the store. B&m targets an upscale group of people who consume most of the premium beverages in the United States. The marketing strategy is based on a combination of selection, price, service, entertainment, and ambience.

INVENTORY AND DISPLAY MANAGEMENT

B&m is concerned about managing inventory so that turnover is sufficient to generate profits. One management tool is a system that tracks the movement of every SKU in every store. The system scans everything that comes in the back and everything that goes out the front, and it produces what is called a sludge report. Wine sellers do not believe that they sell sludge, but as far as B&m is concerned, any wine that does not move becomes sludge. If a wine is not moving, it is put on display at a lower price on a store-by-store basis until it is sold.

The stores are online, with a central computer that allows any one of them to search the total inventory to locate an out-of-stock item. The requesting store can then call to arrange a pickup for, or a delivery to, its customer. The stores do not write orders. The central computer reviews each store's inventory every night to determine SKU movement and writes purchase orders for all the stores. Replenishment buyers review the machine-written purchase orders. If there are no problems, the orders are sent out either online or by faxes that are actually sent electronically.

The stores have tremendous display space. They use free-standing displays called buildups, which are on portable platforms so they can be moved from one location to another. A lot of volume is done from displays of special items selected for promotion. The displays are changed regularly to give repeat customers something new to find. Some special displays stay up all the time, but the products they display change. For example, there is always a display of port wines, critics' choice wines, and cabernets. The critics' choice display presents wines that are highly rated by wine critics or by qualified B&m wine buyers who are experienced wine judges. The products change as new wines are judged.

B&m does not have a central distribution center. In its original concept, the economic model for the company included a warehouse because that is the way business has traditionally been done. But having a warehouse is expensive and, upon reflection, the company decided to invest in electronic inventory management systems, linked to their suppliers, that did away with the need for a central warehouse. By combining superior technology, high-volume stores, and reliable distributors, the company could be organized without a warehouse. This lowered the investment required to start the company and allowed management to focus on buying, merchandising, and operating rather than on managing a distribution business. B&m built a successful business without a distribution center, which helps make it more profitable than much of its competition.

MANAGEMENT AND MARKETING

The store managers are responsible for three activities: taking care of their customers, maintaining high store standards, and training and developing their associates. They are compensated based on store sales and profitability, but they are not buyers and they do not set prices. They tell the buyers about any requests they receive, and they can influence what is carried and displayed, but they do not buy products. To that extent, B&m is very centralized, with regional buyers in Southern California and Florida. Regional buyers are specialists in their region.

B&m has wine and beer tastings in the stores on Fridays and Saturdays and cooking demonstrations on Saturdays. B&m staff chefs or guest chefs from restaurants or wineries may conduct the demonstrations. The staff chefs develop a lot of recipes and train other staff members. This activity is part of the entertainment function of the stores. The idea is to have people come in, on the weekends in particular, and enjoy food, wine, and friendship.

B&m is price oriented but does not necessarily have the lowest prices in the market. The policy is to offer a very good value to customers and to shout about it. The company does price comparisons and shows the results in print ads and on store signs and talks about them on the radio. This alerts consumers (and competitors) to the fact that B&m offers a broad array of wines and other products and is value oriented and not just price oriented.

The company uses newspaper advertising about 12 or 13 weeks per year. This is primarily to exploit marketing opportunities during holiday periods, when interest in wine is high. More frequent print advertising does not pay off, given the relatively low number of frequent wine buyers. This is contrary to the practice of supermarkets and most liquor stores, which tend to use print media throughout the year. B&m uses radio more often, for about 35 weeks in a new market and 26 weeks in an established market where store awareness is high. Radio is a more efficient and targeted medium than newspapers.

Most of B&m's marketing dollars are allocated to direct mail. An important element of that effort is the frequent buyer program called Club Bev, which gives discounts to frequent buyers and provides rich data for the direct mail program. The buyer program is free and provides members a card with a barcode on the back, which allows the company to track the purchases of each member. B&m tracks the purchases of approximately 55 percent of the sales generated. This makes it possible to discover the buying practices and preferences of individual members and fashion direct mail appeals that respond to those patterns. The company thought, for example, that cigar buyers would be primarily consumers of cognac and port. The Club data revealed, however, that there is a much stronger correlation between cigar smoking and consumption of premium chardonnay and cabernet than there is with consumption of port, cognac, and malt whiskey.

If Bob, for instance, is a Club Bev member, company experts will analyze what he purchased, when he purchased it, and where he purchased it and try to figure out how they can affect Bob's shopping frequency and product choices. Sending him information that seems relevant, given his purchasing patterns, would do this. If Bob buys nothing but chardonnay, the company would never send information about new cabernets but would send information about new chardonnays. The company could run an affinity analysis to figure out a companion product for chardonnay, then send Bob information about that companion product.

Database-driven, consumer-specific marketing is one of the most important trends in retailing. A cost-effective marketer delivers information to only the people who are truly interested in it. This concept, implemented through the frequent buyer program and tremendous investment in management systems, differentiates B&m from other retailers and is one of the major reasons for its success.

B&m buys directly from a couple of hundred wineries in California, most of them small. It also buys from distributors and brokers. In Florida, everything comes through distributors. The choice depends on legal regulations and on how wineries want to do business. The company can deal with small quantities from a multitude of small suppliers. Of course, the greatest volume comes from the large producers that dominate the wine market.

This practice is different from that of a supermarket. It is difficult for a small winery, or any other, to sell wine without a Universal Product Code (UPC), and it is difficult to sell in small quantities. Supermarkets are selective in their inventory and cannot deal with small lots that may lead to stockouts. B&m prefers but does not always require a UPC. The fact of life is that in the popular varieties like chardonnay and cabernet, the company has to drop an item in order to take a new item into stock. Thus a vendor should be prepared to argue why product A should supplant product B or product C on the shelf. The really smart vendors say "My product is going to sell better than that one. Look at your data. Here's why mine will sell better. Kick that one out, and put mine in." That is how vendors work with B&m buyers. Sales are easier for wineries selling in categories that are expanding, such as really good pinot noir, petite sirah, and other non-traditional red varietals.

There are some products that must be stocked. When Gallo came out with its new Sonoma line, fully supported by large-scale advertising, there was an immediate need to have it in the stores. Its zinfandel just won a double gold at a major competition. It is a great product with lots of marketing support. B&m has to carry it. However, the buyer's decision is more difficult with a small winery with no marketing program or presence and without a writeup somewhere. Although the sale may be difficult, often a buyer will take a chance. If the buyer thinks the wine is really good and expects to be able to convince the store associates and customers that it is a terrific find, the buyer will buy the wine and push it. The company will work with small wineries, distilleries, and breweries to help build brands that it thinks have a future. The buyers work hard to find the undiscovered wines, put them into the stores, and help build the brands. In its support for small enterprise, B&m is different from most retailers.

Sales in the stores are supported by three important activities. One is the Club Bev program and its mailings of newsletters, postcards, information on products, and offers of discounts. Everyone in the program gets the newsletter, which has discounts for everyone, but there are also discounts that are for specific members. It may be just 2,000 or 3,000 people that would get a deal on $7 zinfandel because they have bought it in the past. The second is the signage in the stores, which is well designed and value focused. The signage ties in directly with the numerous displays of value-oriented products. These include house-label wines bought from certain wineries and produced to the company's preferences. Third, there is always someone in the store who really knows the product. If a customer asks to see the best $7 Riesling, the associate will find it right away. The associates develop relationships with a lot of customers, just as in a fine wine shop.

ESTABLISHING A WINE-SELLING BUSINESS

There are two concepts for establishing a business. One is to copy what others are doing but do it better, and the other is to find a need that others are not filling and fill it. B&m followed the latter approach. It sought to determine if there was a need for the kind of store that it envisioned. To answer that question, B&m analyzed what influences people's choices of places to buy beverages. Research confirmed intuition in concluding that where people bought beverages was a function of the kinds of beverages they bought and how often they bought them.

An infrequent wine buyer is primarily influenced by convenience and is most likely to buy wine at a supermarket while shopping for food. These buyers are not a feasible target for specialty beverage stores. The motivating factors begin to change for a person who drinks wine moderately often. Selection and service become important, and convenience becomes less of a consideration. Selection becomes extremely important for a frequent wine buyer, good value is probably the second consideration, and service is the third consideration but is still important. Different ratings apply to regular or frequent beer drinkers, depending on whether the preference is for popularly priced beer or for craft brews and imports. Price is the most important consideration for popularly priced beer; drinkers of this type of beer will buy it on sale at the supermarket, at the club, or at B&m. Consumers of craft brews and imports rank selection first, price second, and service third. They behave a lot like wine consumers.

These differences in behavior provided an opportunity for B&m to examine how it might serve frequent and moderate beverage users, those buyers motivated by more than just convenience. A price/selection matrix was useful to show where the major wine sellers had positioned themselves. This matrix is illustrated in Exhibit 28–1.

Information developed through the competitive matrix was important in guiding B&m's initial marketing strategies. It helped in the positioning decision and in the subsequent marketing activity decisions. The positions in the matrix are approximate but indicate that price clubs have the lowest prices and most limited selection. Warehouse and discount stores also have low prices and low selection. Typical grocery and liquor stores have higher prices, with the latter perhaps having a slightly larger selection. Wine shops and clubs have excellent selection and high prices. Supermarkets have somewhat lower prices and selections that range from poor to excellent but are generally good. Without B&m the low-price/high-selec-

---

Exhibit 28–1 The Competitive Matrix for Wine Price and Selection

High price	Grocery stores	Wine shops
	Liquor stores	Wine clubs
	Supermarkets	
Low price	Warehouse stores	
	Discount stores	Beverages and more!
	Price clubs	
	Low selection	**High selection**

tion quadrant would have been vacant, as it was when B&m was developing its positioning strategy. That was the opportunity that motivated the establishment of the company.

Costco and Sam's are examples of price club stores. The Wine Exchange and Wine Club are more narrowly focused than B&m. Wine Club carries a couple of malt whiskeys and is not really in the spirits business; it is in the wine business and serves customers to whom wine is extremely important. Such customers shop the Internet, search the catalogs, and look at all the ads. They spend a lot of money. B&m's target market is slightly downscale from this and has a larger sales potential, although there is considerable overlap between target markets of the stores.

A further ranking of beverage outlets was made for performance in specialty food selection, service, and entertainment. The entertainment rating reflects the need for activities that draw consumers into the store. Such activities include food demonstrations, wine and beer tastings, clever signage, and friendly sales associates that take good care of customers. Entertainment in most stores is not necessary because customers go to the stores for required shopping. However, entertainment helps in stores that consumers do not have to patronize. Often B&m customers go to the store to have a good time: they go for cooking demonstrations; to taste good food and wine; or to find a new malt whiskey, new craft brew, or award-winning wine. This entertainment strategy was adopted because it sets B&m apart from many competitors.

The concept that emerged from matrix analysis was pretty straightforward. B&m's objective was to become a category-dominant beverage retailer offering an extensive selection at low prices and a pleasant shopping experience. Prices are 11 percent to 13 percent below those of supermarkets and drugstores on a broad market basket of products month in and month out. Club store prices are lower, but the stores do not offer great selection or service and entertainment. Most competitors do not even attempt to enliven the shopping experience.

Marketers must look beyond demographic data to find opportunities. Changes in the population's mix of ages, income, and education lead to changes in consumer behavior. However, changes in lifestyle are far more important and difficult to evaluate. The decision to establish B&m depended partly on emerging trends in consumer behavior. Behind the raw data about market size and growth were some underlying changes in behavior that indicated that luxury goods and conspicuous consumption were coming into favor. The economy was improving, confidence was growing, and, once again, Americans were saying, "It's OK to have fun." B&m believed that it was on the cutting edge of this change and would be able to capitalize on it. This would not have happened if the bankers and analysts had relied only on demographic data, the mainstay of so many data banks.

The steps in setting up and organizing a business can be condensed into four categories: organization, systems, tests, and growth. The first step for B&m was to build the organization. As founder of the company, I had extensive experience in food and wine retailing and management. The first employee hired was a chief financial officer, followed by people experienced in starting up beverage businesses and doing rapid product rollouts. The second step was to invest heavily in systems. A business thrives and grows on data and the information derived from them. The systems allowed tremendous control over inventories, sales, margins, expenses, and profits. The third step was to test the systems and various strategies through a limited-scale operation. The six stores put into operation in 1994 in the San Francisco Bay Area provided the buying clout and economies of scale needed to prove the model. Stores were improved and practices refined during 1995, and plans for expansion were reviewed against this experience. This period allowed the consolidation of the Bay Area market and relationship building with suppliers so that the company could gain a respectable market share. The fourth step was to launch the expansion that is currently underway. The company reached initially to the San Diego and Los Angeles markets, then to Florida, a large market with less competition than in California.

The difficult part of expansion is not finding real estate but hiring good people. The hardest thing to do is to maintain cultures and standards and hire good people as the firm grows rapidly. This is particularly tough in the beverage business because there are not a lot of people who know about wine and are looking for work. The first attempt, of course, is to hire people who know about the retail business, wine, and sales. The Florida store opened with five designated sellers who met those criteria. Additionally, B&m has a mandatory training program, which requires that within the first 90 days every associate, including associates in the office, pass an in-house course in product knowledge and selling techniques for wine, beer, spirits, and cigars. Associates have to pass the course in order to remain employed. Beyond that there are tastings in the stores every week and advanced training programs in niche areas such as malt whiskey, port, and Bordeaux.

CONCLUSIONS

The underlying theme in this case study is the need for careful examination of consumer buying preferences. This is commonly thought of in connection with sales strategies but is equally important to establishing a business. The use of a competition matrix building on knowledge of a market segment's preferences, in this case the preference for convenience, permits an analysis of competitive positioning and the identification of positioning opportunities for the new or changing firm. Studying this matrix led the firm's founders to successfully create a store that is substantially different from competitor stores—just the way a successful wine is differentiated from competitor wines.

The case study reinforces a marketing maxim that one must look beyond consumer demographics to the factors that make market segmentation useful. In this case, it was not the income class of the consumers that was important but rather their drinking patterns and shopping preferences. This knowledge was used to derive a sales strategy in which a frequent buyer program was developed, store signs were used effectively to stimulate sales, and sales associates were trained to be knowledgeable about the product and skilled in selling it.

The adoption of data-driven marketing strategies is an important trend in the beverage business. These strategies depend on systems that allow quick identification of consumer trends, continual control of inventory and ordering, and rapid responses to changes in financial factors. Club Bev, which allows accurate tracking of customer purchase patterns, is a good example of this type of strategy.

Finally, the four steps in building an organization apply equally well to any marketing strategy. Entrepreneurs must build an organization to establish the strategy, invest in systems to implement and control it, test the results along the way, and then grow. This chapter has illustrated how a successful retailer has applied marketing principles to sell wine and how wineries might apply similar principles in supplying it.

CHAPTER 29

The New Brand in a Competitive Market

Michael C. Houlihan

INTRODUCTION

Barefoot Cellars was organized from day one as a "virtual" winery. That is, it does not have wine making, bottling, or shipping facilities. These activities are contracted to others. The founders realized from the outset that the most important requirement for success in the wine business was not a winery, a vineyard, or even a great wine maker; it was a purchase order. All the rest could be contracted out to others. Almost as important as the purchase order was the ability to merchandise the wines after they were delivered to the buyer. The founders knew that if the wines were not replaced on the shelf immediately after being sold, they would be discontinued—no matter how good the wines were or how fast they sold the first time. This chapter emphasizes the marketing decisions that made Barefoot Cellars a success.

SELECTING THE MARKETING NICHE

Barefoot's founders were successful because they put their money into merchandising and not into bricks, mortar, and vineyards. They hired their own merchandisers to keep the distribution channels flowing. This kept the cash flowing so they could build their virtual winery from 5,000 cases

---

Michael C. Houlihan is a native Californian with extensive experience as an entrepreneur, serving as a business consultant to the wine industry since 1983. In 1986 he founded Barefoot Cellars, of which he is the CEO and president.

per year to 200,000 over a 10-year period. Working with distributors, they realized that their products were a very small percentage of what the distributors' salespeople had to sell. Therefore, in order not to get lost in a distributor's book, they had their merchandising sales representatives "ride with" the distributor's salespeople to make presentations to buyers and to alert them to runout situations. Barefoot would not have had the funds to do this if its money had been tied up in a winery and vineyards. The founders capitalized on the fact that many wineries needed bottling contracts, either because their own products were not selling fast enough to keep their bottling line running, or because they were actually in the business of offering bottling services.

At the outset, the founders made a decision about what kind of wine product they would produce. The owners realized that it was critical to identify a good niche—a well-defined market with great potential for growth. The chosen niche was personal house wine—wine that middle-class people, who may have limited knowledge of wine, use as everyday drinking wine. Personal house wine is often the only wine that these consumers buy. These buyers are the primary, but not the only, targets of Barefoot's marketing efforts.

PRICING

Competitive price points are critical to the success of a personal house wine. Most costs other than the wine itself (such as taxes, bottles, labels, corks, foils, packaging, marketing, and merchandising) are quite similar throughout the industry. A small difference in the largest variable, wine cost, results in a large difference in the shelf price. This suggests that there should be a different approach to positioning a personal house wine. Many wineries figure out what a case of wine has cost them and then determine a price, but Barefoot's founders decided to do the opposite. They looked at what the price *had to be* on the shelf, then figured out what the retailer and distributor *had to have* in margins, then determined what the winery price *must be*, and then worked backward to discover what the wine cost *must be*. It also found what an increase in wine costs could do to the retail price. For example, if increased grape costs forced the cost of wine to increase by 5 percent, this would result in a 50-cent increase to the winery but a $1.25 increase at retail, enough to push the wine far beyond its "normal" price point. Because the market is price sensitive, Barefoot buys wines of different appellations and vintages that meet market taste preferences and price point requirements.

Embraced from the start of the company, this approach gave Barefoot the fiscal direction to be a success in competing with the large competitors in this market. It also meant that Barefoot made little or no profit for the first six years until the combination of consumer and trade recognition, and intensive merchandising took hold. During this period, the business grew in production and sales, spreading overhead costs and gaining efficiencies of scale. This experience proved that the intrinsic value of a virtual business is that you pay as you go without the money and costs associated with major fixed assets and debt. The six-year period required a lot of patience by owners and employees as various strategies were tested and modified and tested again. It also took a certain amount of nerve and faith to wait for results. In the end it was worthwhile, and the philosophy of a virtual business paid off.

PACKAGING

One of the other major ingredients to success was packaging. In developing its packaging, Barefoot first went to buyers to ask whether there was room in their wine sets, what the price of the wine should be, how the package should look, what the customer profile was, and what the customers wanted. The conversations with buyers revealed important information for Barefoot's marketing strategies. A personal house wine had to be at a velocity price point, had to be a true varietal with consistent taste, had to use common-language taste and use descriptors, and had to have a colorful package with a symbol that would be readily recognizable and a name customers would not forget. The wines had to be soft and drinkable, approachable, affordable, and available. The interviews revealed that the 1.5-liter size was the "loyalty" size package. When customers buy the 1.5-liter bottle, they are sold on that particular brand, and they will not shop for another unless the price goes up or the taste stops being good.

THE MARKETING PROGRAM

Another major ingredient of success is the marketing program. There are three ways most wine is sold. One is through a winery's tasting room. Very few wineries sell exclusively in this way, although many use it as part of a mixed strategy. The second way to sell is by direct marketing, using mail, a customer club, telephone sales, or the Internet. The seller develops its own clientele and nurtures it throughout the year. The third and most prevalent way is through mass distribution using brokers, distributors, and retailers, or some combination of them. This is the distribution strategy of most wineries, including every major wine company. It is the strategy used by Barefoot. An important aspect of this is recognizing that there are five points of sale: the distributor, the distributor's salesperson, the salesperson's buyer, the buyer's floor person, and the general public. If all five people are not sold on the product, the wine is likely to fail. Each of these people requires personal attention from the winery, particularly the floor clerk responsible for seeing that shelves are stocked. The winery cannot just hand the retailer a good product that was designed exactly like the buyers wanted it and expect it to sell. It has to be merchandised.

Luckily, expensive TV, radio, and magazine display advertising does not really work in the wine business unless the brand's name is already well established and the advertising is done on a continuous basis. But what if the brand is new and not on many retail shelves, and customers do not know what the wines taste like? The most effective way to overcome this is through public tastings and donations. Into the hand of every taster at every public tasting, Barefoot representatives put a list of retailers that sell Barefoot. Furthermore, Barefoot finds out exactly what nonprofit foundations are the most important to each community in which it markets. Barefoot supports the organizations with donations for fund-raising dinners. It promotes the foundations it supports at the store level with an elaborate neck-tag program. These neck tags may be in support of Ducks Unlimited in Fresno, the AIDS Foundation in San Francisco, the Blue Water Task Force in Los Angeles, or the Seattle Center for the Arts in that city. The philosophy is to be a part of each community. The link is reinforced by a toll-free number on each bottle, a Web page, and hundreds of tastings each year. This helps establish a rapport with customers and has resulted in repeat purchases rising to 80 percent of total sales.

THE MISSION

Barefoot is a success because it views itself as a service, not as a product. Its mission is to provide the best personal house wine at the best price. Medals, awards, and recommendations measure best personal house wine.

The best price is the price at which personal house wines sell the fastest. Our experience is that the best price range was $4.99 to $6.99 per 750 milliliters as measured in 1999. Replacement is more important than placement, so distribution and retail channels must remain full and in motion. Money is better spent on merchandising than on hard assets. Customer loyalty is essential and must be earned every day. If the product is not in stores, it does not matter how good it is. It must be there, taste good, be priced right, be noticeable and recognizable, and have a company behind it with which the customer can identify.

CONCLUSIONS

The principles of marketing are solid and well known. Although they are difficult to implement in a fresh and creative way, Barefoot has succeeded in doing just that. From Barefoot's experience, some of the following common-sense slogans emerge:

- In the wine business, make up your mind whether you want to make a statement or make a deposit (seek out either artistic expression or business success).
- If it's not on the floor, it's not in the store (displays get the product noticed).
- Replacement is better than placement (if it's not there, it won't sell).
- Don't make it if you can't sell it.
- If you're not riding, you're sliding (ride with sales representatives, or you will be out of stock).
- If you can't make them smile, don't make them wine (maintain a good sense of humor and remember that customers have to enjoy the whole approach).

Index